Benjamin Peirce

Elementary Treatise on Curves, Functions and Forces

Vol. 1

Benjamin Peirce

Elementary Treatise on Curves, Functions and Forces
Vol. 1

ISBN/EAN: 9783337779177

Printed in Europe, USA, Canada, Australia, Japan

Cover: Foto ©berggeist007 / pixelio.de

More available books at **www.hansebooks.com**

AN

ELEMENTARY TREATISE

ON

CURVES, FUNCTIONS, AND FORCES.

———

VOLUME FIRST;

CONTAINING

ANALYTIC GEOMETRY

AND THE DIFFERENTIAL CALCULUS.

———

By BENJAMIN PEIRCE, A. M.,

University Professor of Mathematics and Natural Philosophy in Harvard University.

———

BOSTON:

JAMES MUNROE AND COMPANY.

M DCCC XLI.

CONTENTS.

BOOK I.

APPLICATION OF ALGEBRA TO GEOMETRY.

CHAPTER VIII.

CHAPTER IX.

BOOK II.

DIFFERENTIAL CALCULUS.

CHAPTER I.

CHAPTER II.

CHAPTER III.

CHAPTER IV.

CHAPTER V.

CHAPTER VI.

CHAPTER VII.

CHAPTER VIII.

CHAPTER IX.

CHAPTER X.

CHAPTER XI.

CHAPTER XII.

BOOK I.

APPLICATION OF ALGEBRA TO GEOMETRY.

CURVES AND FUNCTIONS.

BOOK I.

APPLICATION OF ALGEBRA TO GEOMETRY.

CHAPTER I.

GEOMETRICAL CONSTRUCTION OF ALGEBRAICAL QUANTITIES.

1. In the application of Algebra to Geometry, usually called *Analytic Geometry*, the magnitudes of lines, angles, surfaces, and solids are expressed by means of letters of the alphabet; and each problem, being put into equations by the exercise of ingenuity, is solved by the ordinary processes of Algebra. The algebraical result is finally to be interpreted geometrically; and this geometrical interpretation of an algebraical expression is called the *geometrical construction* of that expression. The geometrical construction of the results is, then, the last operation in the solution of problems; but it is convenient, on account of its simplicity, to begin with the consideration of it. We begin with the easiest cases and proceed to the more difficult ones, and we regard each letter as representing a line, so that

1

every algebraical expression of the first degree will denote a line ; whence it is called linear.

2. *Problem.* To construct $a + b$.

Solution. Take (fig. 1.)
$$AB = a,$$
$$BC = b\,;$$
and we have
$$AC = AB + BC = a + b\,;$$
so that AC is the required value of $a + b$.

3. *Problem.* To construct $a - b$.

Solution. Take (fig. 2.)
$$AB = a,$$
and from B, in the opposite direction,
$$BC = b\,;$$
we have then
$$AC = AB - BC = a - b\,;$$
so that AC is the required result.

4. *Corollary.* If a were zero, the preceding solution would become the same as to take from A (fig. 3.) in the direction AC, opposite to AB,
$$AC = b\,;$$
so that the negative sign would only be indicated by the direction of AC. In order to generalize the preceding construction we must, then, adopt the rule that

The geometrical interpretation of the negative sign is opposite direction.

5. *Problem.* To construct an algebraic expression consisting of a series of letters connected together by the signs $+$ and $-$.

Product and Quotient.	Surface and Solid.

Solution. Collect into one sum, by art. 2, all the letters preceded by $+$, which sum may be denoted by a; and collect into another sum all the letters preceded by $—$, which sum may be denoted by b; and the value of $a — b$ may then be constructed by art. 3.

6. *Corollary.* If a letter is preceded by an integral numerical coefficient, it may be regarded as a letter repeated a number of times equal to this integer.

7. *Problem.* To construct $a\,b$.

Solution. The parallelogram of which the base is a, and the altitude is b, is equal to the product $a\,b$, which accordingly represents a surface; and this conclusion is a general one, that is,

A homogeneous algebraical expression of the second degree represents a surface.

8. *Problem.* To construct $a\,b\,c$.

Solution. The parallelopiped of which the base is the parallelogram $a\,b$, and the altitude is c, is equal to the product $a\,b\,c$, which accordingly represents a solid; and, in general,

A homogeneous algebraical expression of the third degree represents a solid.

9. *Problem.* To construct $\dfrac{a}{b}$.

Solution. Make (fig. 4.) the right angle ABC, take
$$AB = a$$
$$BC = b,$$
and join AC. The angle ACB is, by trigonometry, that angle whose tangent is $\dfrac{a}{b}$.

10. *Corollary.* If we had taken

$$AC = b,$$

the angle ACB would have been the angle, whose sine is $\dfrac{a}{b}$,

and, in general,

A homogeneous algebraical expression, whose degree is zero, represents the sine, tangent, &c. of an angle.

11. *Scholium.* Since no other magnitudes occur in Geometry but angles, lines, surfaces, and solids, *all algebraical quantities which represent geometrical magnitudes must be either of the* 1st, *the* 2d, *the* 3d, *or the zero degree;* and since dissimilar geometrical magnitudes can neither be added together, nor subtracted from each other, these algebraical expressions *must also be homogeneous.*

If, therefore, an algebraical result is obtained, which is not homogeneous, or is of a different degree from those just enumerated; it can only arise from the circumstance, that the geometrical unit of length, being represented algebraically by 1, disappears from all algebraical expressions in which it is either a factor or a divisor. To render these results homogeneous, then, and of any required degree, it is only necessary to restore this divisor or factor which represents unity.

12. *Problem.* *To render a given algebraical expression homogeneous and of any required degree.*

Solution. Introduce 1, *as a factor or divisor, repeated as many times as may be necessary, into every term where it is required.*

13. EXAMPLES.

1. Render $\dfrac{a^2\,b + c + d^2}{e + h^2}$ homogeneous of the 1st degree.

$$Ans. \quad \frac{a^2\,b + (1)^2\,c + 1.\,d^2}{1.\,e + h^2}.$$

2. Render $\dfrac{a + d^3 + h\,e}{l^2 + m^3}$ homogeneous of the 2d degree.

$$Ans. \quad \frac{(1)^4\,a + (1)^2\,d^3 + (1)^3\,h\,e}{1\,.\,l^2 + m^3}.$$

3. Render $\dfrac{a + b\,c + d^5\,h^2}{d\,e - a}$ homogeneous of the 3d degree.

$$Ans. \quad \frac{(1)^6\,a + (1)^5\,b\,c + d^5\,h^2}{(1)^2\,d\,e - (1)^3\,a}.$$

4. Render $\dfrac{a^2 + b}{c - d^2}$ homogeneous of the zero degree.

$$Ans. \quad \frac{a^2 + 1.\,b}{1.\,c - d^2}.$$

5. Render $a\,b$ homogeneous of the 1st degree.

$$Ans. \quad \frac{a\,b}{1}.$$

6. Render $a\,b\,c + d - e^2$ homogeneous of the 1st degree.

$$Ans. \quad \frac{a\,b\,c}{(1)^2} + d - \frac{e^2}{1}.$$

14. *Scholium.* By the preceding process, every fraction, which does not involve radicals, may be reduced to a homogeneous form, in which each term is of the first degree; and, although this form is not always that which leads to the most simple form of construction, its generality gives it a peculiar fitness for the general

1*

purposes of instruction, where the artifices of ingenuity are rather to be avoided than displayed.

15. EXAMPLES.

1. Reduce the fraction of example 1, art. 13, to a homogeneous form, in which each term is of the first degree.

$$Ans. \quad \left(\frac{a^2\,b}{(1)^2} + c + \frac{d^2}{1}\right) \div \left(e + \frac{h^2}{1}\right).$$

2. Reduce the fraction of example 2, art. 13, to a homogeneous form, in which each term is of the first degree.

$$Ans. \quad \left(a + \frac{d^3}{(1)^2} + \frac{h\,e}{1}\right) \div \left(\frac{l^2}{1} + \frac{m^3}{(1)^2}\right).$$

3. Reduce the fraction of example 3, art. 13, to a homogeneous form, in which each term is of the first degree.

$$Ans. \quad \left(a + \frac{b\,c}{1} + \frac{d^5\,h^2}{(1)^6}\right) \div \left(\frac{d\,e}{1} - a\right).$$

4. Reduce the fraction of example 4, art. 13, to a homogeneous form, in which each term is of the first degree.

$$Ans. \quad \left(\frac{a^2}{1} + b\right) \div \left(c - \frac{d^2}{1}\right).$$

16. *Problem. To construct* $\dfrac{a\,b}{c}$.

Solution. We have

$$c : a = b : \frac{a\,b}{c},$$

that is, the given fraction is a fourth proportional to the three lines, *c*, *a*, and *b*.

Find, then, by geometry, a fourth proportional to the

Monomial.

lines c, a, and b; and this fourth proportional is the required result.

17. *Corollary.* The value of $\dfrac{a^2}{c}$ is a third proportional to c and a.

18. *Problem.* To construct any monomial which denotes a line.

Solution. If the monomial is not of the first degree, reduce it to the first degree by art. 12. It is then of the form

$$\frac{a\,b\,c\,d\ldots}{a'\,b'\,c'\ldots} = \frac{a\,b}{a'} \times \frac{c}{b'} \times \frac{d}{c'}\ldots.$$

Construct first $\dfrac{a\,b}{a'}$, and let m be the line which it represents. The given quantity becomes $\dfrac{m\,c}{b'} \times \dfrac{d}{c'}\ldots$

Let, again, $\qquad\qquad m' = \dfrac{m\,c}{b'}$,

also $\qquad\qquad\qquad m'' = \dfrac{m'\,d}{c'}$, &c.

and the last line thus obtained is plainly the required result.

19. EXAMPLES.

1. Construct the line $a\,b$.   *Ans.* $m = \dfrac{a\,b}{1} = a\,b$.

2. Construct the line $a\,b\,c$.   *Ans.* $m = \dfrac{a\,b}{1}$, $m' = \dfrac{m\,c}{1} = a\,b\,c$.

Any expression not involving radicals.

3. Construct the line a^5.   Ans. $m = \dfrac{a^2}{1}$, $m' = \dfrac{m\,a}{1}$,

$$m'' = \frac{m'\,a}{1}, \; m''' = \frac{m''\,a}{1} = a^5.$$

4. Construct the line $\dfrac{a^2\,b}{d^3}$.   Ans. $m = \dfrac{a^2}{d}$, $m' = \dfrac{m\,b}{d}$,

$$m'' = \frac{m'\,.\,1\,.}{d} = \frac{a^2\,b}{d^3}.$$

5. Construct the line $\dfrac{2\,a\,b}{e\,f\,g}$.   Ans. $m = \dfrac{2a\,.\,b}{e}$, $m' = \dfrac{m\,.\,1}{f}$,

$$m'' = \frac{m'\,.\,1}{g} = \frac{2\,a\,b}{e\,f\,g}.$$

6. Construct the line $\dfrac{1}{a}$.   Ans. $m = \dfrac{(1)^2}{a} = \dfrac{1}{a}$.

7. Construct the line $\dfrac{A}{B}$.   Ans. $m = \dfrac{1\,.\,A}{B} = \dfrac{A}{B}$.

20. *Corollary. By this process each term of an algebraic expression, which does not involve radicals, is reduced to a line; and if the expression does not involve fractions, it may then be reduced to a single line by art. 5; if it does involve fractions, the numerator and denominator of each fraction is, by art. 5, reduced to a single line, and each fraction, being then of the form* $\dfrac{A}{B}$, *is constructed like example 7 of the preceding article, and the aggregate of the fractions is then reduced to a single line, by art. 5.*

Any algebraic expression, which represents a line, and does not involve radicals, may therefore be constructed by this process.

Any expression free from radicals.

21. Examples.

1. Construct the line $\dfrac{a^2\,b + c + d^2}{e + h^2}$.

Solution. Let $m = \dfrac{a^2\,b}{(1)^2} = a^2\,b$

$$m' = \dfrac{d^2}{1} = d^2$$

$$m'' = \dfrac{h^2}{1} = h^2$$

and the fraction becomes

$$\dfrac{m + c + m'}{e + m''};$$

let now

$$A = m + c + m'$$
$$B = e + m'',$$

and the line represented by $\dfrac{1 \cdot A}{B}$ is the required line.

2. Construct the line represented by the fraction of example 2, art. 13.

> *Ans.* $m' = d^3$, $m'' = h\,e$, $m''' = l^2$, $m^{IV} = m^3$,
> $A = a + m' + m''$, $B = m''' + m^{IV}$,

and the required line is the fourth proportional to B, 1, and A.

3. Construct the line represented by the fraction of example 3, art. 13.

> *Ans.* Let $m = b\,c$, $m' = d^5\,h^2$, $m'' = d\,e$,
> $A = a + m + m'$, $B = m'' - a$,

and the required line is the fourth proportional to B, 1, and A.

4. Construct the line represented by the fraction of example 4, art. 13.

$$Ans. \qquad m = a^2, \quad m' = d^2,$$
$$A = m + b, \quad B = c - m',$$

and the required line is the fourth proportional to B, 1, and A.

5. Construct the line represented by the polymonial of example 6, art. 13.

$$Ans. \qquad m = abc, \quad m' = e^2,$$

and $\qquad A = m + d - m'$ is the required line.

22. Problem. *To construct the line* $\sqrt{(ab)}$.

Solution. Since $\sqrt{(ab)}$ is a mean proportional between a and b, the required result is obtained by *constructing, geometrically, this mean proportional between a and b.*

23. Corollary. The expression

$$\sqrt{A} = \sqrt{(1 . A)}$$

may be constructed by finding a mean proportional between 1 and A.

24. Corollary. *The square root of any algebraical expression, which does not involve radicals, may be constructed by finding, as in art. 19, the line A, which this algebraic expression represents, and then constructing* $\sqrt{A}$ *as in the preceding article.*

By the repeated application of this process, any algebraic expression may be constructed which represents a line, and which does not involve any other radicals than those of the second degree.

Radicals of the second degree.

25. EXAMPLES.

1. Construct the line $\sqrt{(a + b - c - e.)}$
        Ans. $A = a + b - c - e,$
and $\sqrt{A}$ is the required line.

2. Construct the line $\sqrt{(a^2 + a\,m)}$.
        Ans. $A = a^2 + a\,m,$
and $\sqrt{A}$ is the required line.

3 Construct the line $\dfrac{\sqrt{a\,b} + \sqrt{(e\,f - h)}}{e^2 - \sqrt{(c^2 - f\,n)}}.$

Ans. $m = \sqrt{a\,b},\ m' = \sqrt{(e\,f - h)},\ m'' = e^2,\ m''' = \sqrt{(c^2 - f\,n)},$
        $A = m + m',\quad B = m'' - m''',$

and the line $\dfrac{A}{B}$ is the required line.

4. Construct the line $\sqrt[4]{a}$.   *Ans.* $m = \sqrt{a},$
and $\sqrt{m}$ is the required line.

26. *Scholium.* When the expression whose square root is required is easily decomposed into two factors, it is immediately reduced to the form $\sqrt{(a\,b)}$ and constructed, as in art. 22.

27. EXAMPLES.

1. Construct example 2, art. 25, by decomposing the quantity under the radical sign into two factors.

Solution. $a^2 + a\,m = a\,(a + m).$

Let $b = a + m,$
and the line $\sqrt{(a\,b)}$ is the required line.

Square root of sum and difference of squares.

2. Construct $\sqrt{(a^2 + ae + am - an)}$ by decomposing the quantity under the radical sign into two factors.

$$Ans. \quad b = a + e + m - n,$$

and $\sqrt{(ab)}$ is the required line.

3. Construct $\sqrt{(a + a^2 - a^3)}$ by decomposing the quantity under the radical sign into two factors.

$$Ans. \quad b = 1 + a - a^2,$$

and $\sqrt{(ab)}$ is the required line.

4. Construct $\sqrt{(a^2 - b^2)}$ by decomposing the quantity under the radical sign into factors.

$$Ans. \quad c = a + b, \quad e = a - b,$$

and $\sqrt{(ce)}$ is the required line.

28. *Scholium.* Example 4 of the preceding article may also be solved by *constructing a right triangle, of which a is the hypothenuse and b a leg, and* $\sqrt{(a^2 - b^2)}$ *will be the other leg.*

29. *Corollary.* In the same way $\sqrt{(a^2 + b^2)}$ *is the hypothenuse of a right triangle, of which a and b are the legs.*

30. *Corollary. By combining the processes of the two preceding articles, any such expression as*

$$\sqrt{(a^2 + b^2 - c^2 - e^2 + h^2 +)}, \&c.$$

may be constructed. For if we take

$$m = \sqrt{(a^2 + b^2)}, \quad m' = \sqrt{(m^2 - c^2)},$$

$$m'' = \sqrt{(m'^2 - e^2)}, \quad m''' = \sqrt{(m''^2 + h^2)}, \&c.$$

we have $\qquad m^2 = a^2 + b^2,$

$$m' = \sqrt{(m^2 - c^2)} = \sqrt{(a^2 + b^2 - c^2)},$$

or $\qquad m'^2 = a^2 + b^2 - c^2;$

Construction of radicals.

$$m'' = \sqrt{(m'^2 - e^2)} = \sqrt{(a^2 + b^2 - c^2 - e^2)},$$

or $$m''^2 = a^2 + b^2 - c^2 - e^2;$$

$$m''' = \sqrt{(m''^2 + h^2)} = \sqrt{(a^2 + b^2 - c^2 - e^2 + h^2)}, \&c.$$

31. *Corollary. The square root of the sum or difference of any expressions, which involve no other radicals than those of the second degree, may also be constructed by the preceding process. For if either of these expressions is constructed by the processes before given, it may be represented by A; and, if we denote $\sqrt{A}$ by m, we have*

$$m^2 = A,$$

so that each expression is reduced to the form of a square, and the whole radical is reduced to the form of the preceding article.

32. EXAMPLES.

1. Construct example 1, art. 24, by the process of art. 30.
    Ans. $m = \sqrt{a}$, $m' = \sqrt{b}$, $m'' = \sqrt{c}$, $m''' = \sqrt{e}$,
and the line $\sqrt{(m^2 + m'^2 - m''^2 - m'''^2)}$ is the required line.

2. Construct example 2, art. 24, by the process of art. 30.
    Ans. $m' = \sqrt{(a\,m)}$, and $\sqrt{(a^2 + m'^2)}$ is the required
       line.

3. Construct the line $\sqrt{(a^2 + b\,c - e^3 + h)}$ by process of art. 30.
    Ans. $m = \sqrt{(b\,c)}$, $m' = \sqrt{e^3}$, $m'' = \sqrt{h}$,
and the line $\sqrt{(a^2 + m^2 - m'^2 + m''^2)}$ is the required line.

33. *Problem. To construct an algebraical expression which represents a surface.*

2

Surface.	Solid.	Angle.

Solution. Let A be the line which is represented by this algebraical expression, and since we have

$$A = 1 \cdot A,$$

the required surface is represented in magnitude by the parallelogram, whose base is 1, and altitude A, or by the equivalent square, triangle, &c.

34. *Problem.* To construct an algebraical expression which represents a solid.

Solution. Let A be the line which is represented by this expression, and since we have

$$A = (1)^2 \cdot A,$$

the required solid is represented in magnitude by the parallelopiped, whose base is the square $(1)^2$, and whose altitude is A.

35. *Problem.* To construct an algebraical expression, which represents the sine, tangent, &c. of an angle.

Solution. Let A be the line which is represented by this expression, and since we have

$$A = \frac{A}{1},$$

the required angle is found by art. 9 or 10.

36. *Scholium.* The construction of all geometrical magnitudes being, by the three preceding articles, reduced to that of the line; we shall limit our constructions hereafter to that of the line.

37. *Problem.* *To construct the root of an equation of the first degree with one unknown quantity.*

Solution. *Every equation of the first degree may, as is proved in Algebra, be reduced to the form*

$$A x + M = 0 ;$$

whence

$$x = - \frac{M}{A} ;$$

and this value of x may be constructed by art. 18.

38. *Problem.* *To construct the roots of an equation of the second degree with one unknown quantity.*

Solution. *The equation of the second degree may, as is shown in Algebra, be reduced to the form*

$$A x^2 + B x + M = 0.$$

If we divide this equation by A, and put

$$a = \frac{B}{2 A}, \quad m = \frac{M}{A},$$

it becomes

$$x^2 + 2 a x + m = 0.$$

The roots of this last equation are

$$x = - a \pm \sqrt{(a^2 - m)}.$$

Case 1. *When m is positive and greater than a^2, the roots are both imaginary, and cannot be constructed.*

Case 2. *When m is positive and equal to a^2, each root is equal to $- a$, which needs no farther construction.*

Case 3. *When m is positive and less than a^2. Let, in this case,* $b = \sqrt{m}$, *or* $b^2 = m$.

The roots become
$$x = -a \pm \sqrt{(a^2 - b^2)},$$
which are thus constructed.

Draw (fig. 5.) *the two indefinite lines DAD' and AB perpendicular to each other. Take*
$$AB = b\,;$$
from B as a centre, with a radius
$$BC = a,$$
describe an arc cutting DAD' in C. Take
$$CD = CD' = BC = a,$$
and the required roots, independently of their signs, are
$$AD \text{ and } AD'.$$

Demonstration. For
$$AC = \sqrt{(BC^2 - AB^2)} = \sqrt{(a^2 - b^2)}$$
and $-AD = -CD + AC = -a + \sqrt{(a^2 - b^2)}$
$$-AD' = -CD' - AC = -a - \sqrt{(a^2 - b^2)}.$$

Case 4. *When m is zero, the roots are*
$$x = 0 \text{ and } x = -2a,$$
which require no further construction.

Case 5. *When m is negative, so that $-m$ is positive. Let* $b = \sqrt{-m}$, *or* $b^2 = -m$.

The roots become
$$x = -a \pm \sqrt{(a^2 + b^2)};$$
which are thus constructed.

Draw (fig. 6.) *the two lines AB and AC perpendicular to each other.*   *Take*

$$AB = a, \text{ and } AC = b;$$

through BC draw the indefinite line BCD'.   *Take*

$$BD = BD' = AB = a,$$

and the required roots, independently of their signs, are CD and CD'.

Demonstration.   For

$$BC = \sqrt{(AB^2 + AC^2)} = \sqrt{(a^2 + b^2)}$$

and

$$CD = - BD + BC = - a + \sqrt{(a^2 + b^2)}$$
$$- CD' = - BD' - BC = - a - \sqrt{(a^2 + b^2)}.$$

39. *Scholium.* Radicals of a higher than the second degree, and roots of equations of a higher than the second degree, do not usually admit of geometrical construction.

CHAPTER II.

ANALYSIS OF DETERMINATE PROBLEMS.

40. Geometrical problems are of two classes, *determinate* and *indeterminate*.

Determinate problems are those, which lead to as many algebraical equations as unknown quantities; and *indeterminate* problems are those, in which the number of equations is less than that of the unknown quantities.

41. The solution of a geometrical problem consists of these three parts ;

First, the putting of the question into equations ;

Secondly, the solution of these equations ;

Thirdly, the geometrical construction of the algebraical results.

The last of these processes has been treated of in the preceding chapter, but it must be observed that much skill is often shown in arranging the construction in such a form, that it may be readily drawn and be neat in its appearance.

The second process is exclusively algebraical, and the first process, the putting into equations, is a task which, necessarily, requires ingenuity, and can only be taught by examples. One great object is to obtain the simplest possible equations, and such as do not surpass the second degree. It is not unfrequently the case, that, when a question admits of several solutions, two or more of these solutions are connected together in such a way, that the same quantity, being obviously common to them, should, on this account, be selected as the unknown quantity.

42. EXAMPLES.

1. To divide a line AB (fig. 7.) into two such parts, that the difference of the squares described upon the two parts may be equal to a given surface.

Solution. Let the magnitude of the given surface be equal to that of the square whose side is AC, and let D be the point of division, AD being the greater part. Let

$$a = AB,\ b = AC,\ \text{and}\ x = AD,$$

we have then

$$BD = a - x;$$

and the equation for solution is

$$x^2 - (a - x)^2 = b^2,$$

or

$$2\,a\,x - a^2 = b^2.$$

Hence

$$x = \frac{b^2 + a^2}{2\,a} = \frac{b^2}{2a} + \tfrac{1}{2}\,a.$$

Construction. Let E be the middle of AE. Draw the indefinite line EB'. Take

$$EB' = EB = \tfrac{1}{2}\,a,$$

$$EC = EC'' = \tfrac{1}{2}\,AC = \tfrac{1}{2}\,b.$$

Join $B'\,C$, and through C'' draw $C''\,D$ parallel to $B'\,C'$, D is the point of division required.

Demonstration. We have

$$EB' : EC = EC'' : ED,$$

or

$$\tfrac{1}{2}\,a : \tfrac{1}{2}\,b = \tfrac{1}{2}\,b\ : ED;$$

whence $$ED = \tfrac{1}{4}\, b^2 \div \tfrac{1}{2}\, a = \frac{b^2}{2\,a},$$

and $$AD = ED + AE = \frac{b^2}{2\,a} + \tfrac{1}{2}\, a.$$

2. To inscribe in a triangle ABC (fig. 8.), a rectangle $DEFG$ whose base and altitude are in the given ratio $m : n$.

Solution. Let fall the perpendicular AIH. Let

$$BC = b, \quad AH = h,$$

$$DE = HI = x, \quad AI = AH - HI = h - x;$$

and, since $$n : m = DE : EF,$$

we have $$EF = \frac{m\,x}{n}.$$

But the triangles AEF and ABC are similar, and their bases are, therefore, proportional to their altitudes, that is,

$$BC : EF = AH : AI,$$

or $$b : \frac{m\,x}{n} = h : h - x.$$

Hence we find, by algebraical solution,

$$x = \frac{n\,b\,h}{m\,h + n\,b} = b\,h \div \left(\frac{m\,h}{n} + b \right).$$

Construction. Find a fourth proportional to n, m, and h, and denote it by h', and then x is obviously a fourth proportional to $h' + b$, b and h.

The following simple form has been obtained by geometers. Draw AK parallel to BC, and take

$$AK = h'.$$

Join KC, and ED is the required altitude.

Line of given length intercepted between parallels.

Demonstration. For since h', or its equal AK, is a fourth proportional to n, m, and h, we have

$$n : m = h : AK,$$

or $$AK : AH = m : n.$$

If we let fall the perpendicular KL upon BC, we have the quadrilateral $CFED$, $CAKL$, which are formed of similar triangles; they are therefore similar, and their homologous sides give the proportion

$$FE : DE = AK : KL \text{ (or } AH) = m : n;$$

so that FE and DE are in the required ratio.

Corollary. If the ratio $m : n$ were that of equality, the rectangle would be a square.

3. To draw through a given point A (fig. 9.) situated between two given parallels BC and DE a line HI, which may be of a given length a.

Solution. Since the point is given, its distances from the parallels must be given, which are

$$AF = b, \quad AG = c;$$

let $$AH = x;$$

we shall leave it as an exercise for the learner to find the value of x, which is

$$x = \frac{a\,b}{b + c}.$$

Construction. The value of x is a fourth proportional to $b + c$, a, and b, and may be easily constructed.

The following form is quite simple. From G as a centre, with a radius equal to a, describe an arc cutting BC in K. Join GK, and the line drawn through A parallel to GK is obviously of the same length with GK, and it is, therefore, the required line.

Corollary. The problem is impossible when the length a is less than GF or its equal $b + c$.

4. To draw a circle through two given points A and B (fig. 10.), and tangent to a given line DC.

Solution. Join AB, and produce AB to meet DC at D. Let C be the point of contact, and let

$$DA = a, \quad DB = b, \quad DC = x';$$

we have, by geometry,

$$DA : DC = DC : DB,$$

or $\qquad a : x = x : b;$

whence $\qquad x = \pm \sqrt{(ab)}.$

Construction. Find a mean proportional between a and b, and take DC or DC' equal to it, and C or C' is the point of contact, these two values corresponding to the two different circles BCA and $BC'A$.

Instead of finding the mean proportional by the ordinary process, we may find it, by drawing any arc AEB through A and B, and the tangent DE to this arc is, by geometry, the mean proportional between DA and DB.

Corollary. The problem is impossible if a and b are of opposite signs, that is, if A and B are in opposite directions from D, one being above the line and the other below it.

Corollary. If either a or b is zero, as in fig. 11, where

$$DA = a = 0,$$

the problem is reduced to that of finding a circle which passes through the given point B, and is tangent to a given line CA at a given point A.

Construction of this case.   Erect OA perpendicular to AC. Join AB, and at the middle E of AB erect the perpendicular

---

Division of a line.

EO; O is the centre. The demonstration of this construction is left as an exercise for the learner.

Corollary. If a and b are equal, as in fig. 12, the problem becomes; to find a circle which touches a given line DA at a point A, and also touches another given line DC.

Construction. Take

$$DC = DC' = a,$$

and the point O or O', the intersection of the perpendicular OAO', with the perpendicular CO or $C'O'$, is the centre of the required circle.

5. To divide a given line AB (fig. 13.) into two such parts, that the sum of the squares described upon the two parts may be equal to a given surface.

Solution. Let the given surface be twice the square whose side is AC, and let D be the point of division. Let also E be the middle of the line, and let

$$BE = AE = a, \ AC = b, \ DE = x.$$

The value of x will be found to be

$$x = \pm \sqrt{(b^2 - a^2)},$$

so that x is a leg of a right triangle whose hypothenuse is b and other leg a.

Corollary. The problem is impossible when b is less than a and also when

$$x > a,$$

or

$$b^2 - a^2 > a^2,$$

or

$$b^2 > 2\,a^2,$$

or

$$2\,b^2 > 4\,a^2,$$

$$2\,b^2 > (2\,a)^2 \,;$$

that is, when the given surface is greater than the square of the given line.

6. To divide a given line AB (fig. 14.) at the point C in extreme and mean ratio.

Solution. Let AC be the greater part, and let
$$AB = a, \quad AC = x, \quad CB = a - x,$$
we are to have
$$a : x = x : a - x,$$
whence we find

$x^2 + a x - a^2 = 0$, and $x = \frac{1}{2} a \left(-1 \pm \sqrt{5}\right)$.

Construction. The roots of the equation
$$x^2 + a x - a^2 = 0,$$
being constructed by case 5, art. 38, give the usual construction of this problem.

7. Through a given point C (fig. 15.) to draw a line BCD, so that the surface of the triangle ABD intercepted between the lines AB and AD may be of a given magnitude.

Solution. Let the given surface be double that of the given rhombus $AEFG$. Draw CH parallel to AD, and CI parallel to AB. Let
$$AI = CH = a, \quad AH = CI = b,$$
$$AE = c, \quad AD = x, \quad AB = y.$$
We have
$$\text{surface of triangle} = \tfrac{1}{2} x y \sin. A = 2 c^2 \sin. A,$$
whence $\qquad\qquad x y = 4 c^2.$

The similar triangles BHC, BAD, give
$$BH : HC = BA : AD,$$
or $\qquad\qquad y - b : a = y : x;$

whence $\qquad\qquad x y = a y + b x.$

The solution of these equations gives

$$x = \frac{2c}{b}\left(c \pm \sqrt{(c^2 - ab)}\right)$$

$$y = \frac{2c}{a}\left(c \mp \sqrt{(c^2 - ab)}\right)$$

which are easily constructed.

Corollary. The problem is impossible when c^2 is less than ab; that is, when

$$c^2 \sin. A < ab \sin. A,$$

or when the rhombus $AEFG$ is less than the parallelogram $AHCI$.

8. Through a given point C (fig. 16.) to draw a line BCD, so that the part BCD intercepted between two given lines AB and AD may be of a given length, the point C being at equal distances from the two given lines.

Solution. Draw CH and CI parallel respectively to AB and AD, and they are obviously equal to each other. Let

then $AH = AI = CH = CI = a, \quad BD = b,$

$$AD = x, \quad AB = y.$$

From triangle ABD, we have

$$x^2 + y^2 - 2xy \cos. A = b^2;$$

and, from similar triangles BIC and BAD,

$$xy = a(x + y).$$

As these equations are symmetrical with regard to x and y, they are simplified by putting

$$x + y = s, \quad xy = t;$$

3

Given length intercepted.

and become $\qquad t = as$
$$s^2 - 2a(1 + \cos. A)s = b^2;$$
whence s and t, x and y are found.

The following solution is, however, much neater. Join AC, and let the angle ACD be the unknown quantity, and put

$$c \doteq AC, \quad CAD = \tfrac{1}{2}A = A', \quad ACD = \varphi,$$
$$ADC = 180° - (\varphi + A'), \quad ABC = \varphi - A',$$
$$\sin. ADC = \sin. (\varphi + A');$$

and, by trigonometry,

$$\sin. ADC : \sin. DAC = AC : DC,$$
$$\sin. (\varphi + A') : \sin. A' = c : DC;$$

Hence $\qquad DC = \dfrac{c \sin. A'}{\sin.(\varphi + A')}.$

Also $\qquad \sin. ABC : \sin. BAC = AC : BC,$
$$\bullet \sin. (\varphi - A') : \sin. A' = c : BC;$$
$$BC = \frac{c \sin. A'}{\sin. (\varphi - A')};$$

$$b = BD = BC + DC = \frac{c \sin. A'}{\sin. (\varphi - A')} + \frac{c \sin. A'}{\sin.(\varphi + A')};$$

$$b \sin.(\varphi - A') \sin.(\varphi + A') = c \sin. A'[\sin.(\varphi + A') + \sin.(\varphi - A')].$$

But, by trigonometry,

$$\sin. (\varphi + A') = \sin. \varphi \cos. A' + \cos. \varphi \sin. A',$$
$$\sin. (\varphi - A') = \sin. \varphi \cos. A' - \cos. \varphi \sin. A';$$

so that

$$\sin.(\varphi + A') + \sin.(\varphi - A') = 2\sin. \varphi \cos. A';$$
$$\sin. (\varphi - A') \sin. (\varphi + A') = \sin.^2 \varphi \cos. A'^2 - \cos.^2 \varphi \sin.^2 A'$$
$$= \sin.^2 \varphi (1 - \sin.^2 A') - (1 - \sin.^2 \varphi) \sin.^2 A'$$
$$= \sin.^2 \varphi - \sin.^2 A',$$

Given length intercepted.

---

which, substituted in the preceding equation, give

$$b \sin.^2 \varphi - b \sin.^2 A' = 2\,c \sin. \varphi \sin. A' \cos. A';$$

from which we find the value of sin. φ

$$b \sin. \varphi = \sin. A' \left[c \cos. A' \pm \sqrt{(b^2 + c^2 \cos.^2 A')} \right].$$

Corollary. Of the two values of sin. φ, one is clearly negative; and this value corresponds to the line $CB'D'$, which meets AD produced in D', so that

$$CB' + CD' = b.$$

CHAPTER III.

POSITION.

43. As almost all geometrical problems involve more or less the elements of Position, it is important to adopt some convenient method of determining and denoting them. We shall, at first, confine ourselves to the consideration of position in a plane, and then proceed to that of position in space.

44. *Problem.* *To determine and denote the position of points in a plane.*

Solution. The most natural method of determining the position of a point is by its distance and direction ; it is thus, that, if a man wishes to go to any place, he starts in the direction of the place, and proceeds a distance equal to that of the place. Some point as *A* (fig. 17.) must then be fixed upon in the plane to which all the other points *B*, *B'*, &c. may be referred ; and the elements of position of *B*, *B'*, &c., are the distances *AB*, *AB'*, &c., and the angles which *AB*, *AB'*, &c. make with some assumed direction, as that of *AC*, for instance. We shall denote the distances *AB*, *AB'*, &c. by r, r', &c., and the angles *BAC*, *B'AC*, &c by φ, φ', &c.

45. *Definitions.* The point *A*, which is thus fixed upon to determine the other points, is called *the origin of coördinates*, or simply *the origin*.

The line AC is called *the axis of coördinates*, or simply *the axis*.

The distance of a point from the origin is called its *radius vector*, thus r, r' &c. are the radii vectores of B, B', &c.

The radius vector and the angle which it makes with the axis are called *polar coördinates*.

When the position of a point is given, its coördinates must be regarded as given.

Negative radii vectores are entirely avoided by regarding the angles as counted from zero to four right angles. Thus the coördinates of B'' are not the angle CAB and $-AB''$, but they are AB'' and $-CAB''$,

or　　$360° - CAB'' = 180° + C'AB'' = 180° + CAB.$

46. It is often found in the course of a solution, that the origin and axis which have been assumed do not furnish the most simple results; it is desirable, in such a case, to have formulæ by which the elements of position can be readily referred to some other origin and axis.

The referring of the elements of position from one origin and axis to others is called *the transformation of coördinates*.

47. *Problem. To transform coördinates from one system of polar coördinates to another system, which has the same origin but a different axis.*

Solution. Let A (fig. 18.) be the origin, AC the original axis, and AC_1 the new axis. The radius vector is the same in both systems. Let the coördinates of any point, as B, in the first system, be

8*

$$AB = r, \text{ and } BAC = \varphi;$$

and let its coördinates in the new system be

$$AB = r, \text{ and } BAC_1 = \varphi_1;$$

we are to find φ, in terms of φ_1.

Now let $\quad\quad\quad \alpha = CAC_1$

be the angle of the two axes,

we have $\quad\quad\quad BAC = BAC_1 + CAC_1,$

or $\quad\quad\quad\quad \varphi = \varphi_1 + \alpha,$

which is the required formula for transformation.

48. *Problem. To transform coördinates from one system of polar coördinates to any other system.*

Solution. Let A (fig. 19.) be the first origin, and AC the axis; and let A_1 be the new origin, and $A_1 C_1$ the new axis. The coördinates of any point, as B, with reference to the first origin and axis are

$$AB = r, \text{ and } BAC = \varphi;$$

and the new coördinates are

$$A_1B = r_1, \text{ and } BA_1C_1 = \varphi_1;$$

and we are to find r and φ in terms of r_1 and φ_1.

The coördinates of the new origin referred to the first origin and axis must be known; let them be

$$AA_1 = a, \text{ and } A_1AC = \beta;$$

the inclination of two axes must also be known, and let it be α. Produce C_1A_1 to A', we have

$$A_1A'C = \alpha, \text{ and } AA'A_1 = 180° - \alpha;$$

also $\quad AA_1A' = A_1A'C - A_1AA' = \alpha - \beta.$

Distance of two points.

$$AA_1B = 180° - (BA_1C_1 + AA_1A')$$
$$= 180° - (\varphi_1 + \alpha - \beta.)$$
$$\cos. AA_1B = -\cos. (\varphi_1 + \alpha - \beta).$$

The triangle AA_1B gives, then,

$$AB^2 = AA_1^2 + A_1B^2 - 2.AA_1.A_1B.\cos. AA_1B,$$
$$r^2 = a^2 + r_1^2 + 2\,a\,r_1 \cos. (\varphi_1 + \alpha - \beta);$$
$$r = \sqrt{[a^2 + r_1^2 + 2\,a\,r_1 \cos.(\varphi_1 + \alpha - \beta)]} \qquad (1)$$

and $AB : A_1B = \sin. (AA_1B) : \sin. BAA_1,$

or $r : r_1 = \sin. (\varphi_1 + \alpha - \beta) : \sin. (\varphi - \beta);$

whence $\sin. (\varphi - \beta) = \dfrac{r_1}{r}.\sin. (\varphi_1 + \alpha - \beta) \qquad (2)$

and equations (1) and (2) are the required formulæ.

49. Corollary. If the new origin is in the former axis, and if the axes coincide, we have

$$\alpha = 0, \quad \beta = 0,$$

and equations (1) and (2) become

$$r = \sqrt{(a^2 + r_1^2 + 2\,a\,r_1 \cos. \varphi_1)} \qquad (3)$$

$$\sin. \varphi = \frac{r_1}{r}.\sin. \varphi_1. \qquad (4)$$

50. Problem. *To find the distance of two given points from each other.*

Solution. Let B and B' (fig. 20.) be the two points whose coördinates are respectively $r\ \varphi$, and $r'\ \varphi'$. The triangle BAB' gives

$$BB'^2 = r^2 + r'^2 - 2\,r\,r' \cos. (\varphi' - \varphi),$$
$$BB' = \sqrt{[r^2 + r'^2 - 2\,r\,r' \cos. (\varphi' - \varphi)]}. \qquad (5)$$

---

51. *Corollary.* If the point B' is in the axis, we have

$$\varphi' = 0,$$

and (5) becomes

$$BB' = \sqrt{(r^2 + r'^2 - 2\,r\,r'\;\cos.\;\varphi)}. \qquad (6)$$

52. *Corollary.* If B' is the origin, we have

$$r' = 0,$$

and (5) becomes

$$BB' = \sqrt{r^2} = r,$$

as it should be.

53. *Corollary.* If two points are upon the same radius vector, we have

$$\varphi' = \varphi,$$

and (5) becomes

$$BB' = \sqrt{(r^2 + r'^2 - 2\,r\,r')} = r' - r. \qquad (7)$$

54. *Corollary.* If the two points are upon opposite radii vectores, we have

$$\varphi' = \varphi + 180°,$$

and (5) becomes

$$BB' = \sqrt{(r^2 + r'^2 + 2\,r\,r')} = r' + r. \qquad (8)$$

55. Although polar coördinates are the most natural elements of position, they are not those which are usually the most simple in their applications. It has been found convenient to adopt, in their stead, the distances from two axes drawn perpendicular to each other through the origin.

The distances of a point from two axes, drawn perpendicular to each other, are called *rectangular coordinates.*

Thus, if XAX' and YAY' (fig. 21.) are the axes, the rectangular coördinates of the points B, B', &c. are, respectively, BP and BR, $B'P'$ and $B'R'$, &c. We shall denote the distances BR, $B'R'$, &c. from the axis YAY' by x, x', &c., and the distances BP, $B'P'$, &c. from the axis XAX' by y, y', &c.

The distances x, x', &c. may be called *abscissas*, to distinguish them from y, y', &c., which are called *ordinates*.

56. When the rectangular coördinates of a point are known, it is easily found by measuring off its distance x from the axis YAY' upon the axis XAX', and its distance y from the axis XAX' upon the axis YAY', and the lines, which are drawn through the points P and R thus determined, perpendicular to the axes, intersect each other at the required point.

Since the distances x, x', &c. are thus measured upon the axis XAX', this axis is called *the axis of* x, *or the axis of the abscissas;* while the axis YAY' is called *the axis of* y, *or the axis of the ordinates.*

57. By using the negative sign, as in art. 4, the sign of the abscissa, or of the ordinate, designates upon which side of the axis the point is placed.

Thus if we denote, by *positive* ordinates, distances *above* the axis XAX', and by *positive* abscissas, distances *to the right* of the axis YAY', *negative* ordinates will denote distances *below* the axis XAX', and *negative* abscissas, distances to the left of YAY'.

Points in the quarter YAX, being above the axis XAX' and to the right of YAY', will then have positive ordinates and abscissas. Points in the quarter YAX', being above XAX' and to the left of YAY', will have positive ordinates and negative abscissas. Points in the quarter XAY', being below XAX' and to the right of YAY', will have negative ordinates and positive abscissas. Points in the quarter $X'AY'$, being below XAX' and to the left of YAY', will have negative ordinates and abscissas.

58. *Corollary.* For any point in the axis XAX' the ordinate is zero, that is,

$$y = 0$$

is the algebraical condition that a point is in the axis of x.

For any point in the axis YAY', the abscissa is zero, that is,

$$x = 0$$

is the algebraical condition that a point is in the axis of y.

The coördinates of the origin are

$$x = 0, \quad y = 0.$$

59. *Problem.* *To transform from polar to rectangular coördinates.*

Solution. Let A (fig. 22.) be the polar origin, and AC the polar axis. Let A_1 be the new origin, whose position is determined by the coördinates

$$AA_1 = a, \quad A_1AC = \beta.$$

Let A_1X be the axis of abscissas, and A_1Y those of ordinates; and let the inclination of the axis A_1X to AC be α; so that if the line AD is drawn parallel to A_1X, we have

$$\alpha = DAC.$$

Polar transformed to rectangular coördinates.

The values of the polar coördinates

$$AB = r, \text{ and } BAC = \varphi,$$

are to be found in terms of the rectangular coördinates

$$A_1 P = x, \text{ and } BP = y.$$

Produce BP to P', and $A_1 Y$ to A'.   We have

$$A_1 AA' = A_1 AC - A'AC = \beta - \alpha,$$
$$BAA' = BAC - A'AC = \varphi - \alpha.$$

The right triangles $A_1 AA'$ and BAP' give

$$A_1 A' = PP' = AA_1 . \sin. A_1 AA' = a \sin. (\beta - \alpha),$$
$$AA' \qquad = AA_1 . \cos. A_1 AA' = a \cos. (\beta - \alpha) ;$$
$$BP' = BP + PP' = y + a \sin. (\beta - \alpha),$$
$$AP' = P'A' + A'A = x + a \cos. (\beta - \alpha) ;$$
$$AB^2 = (AP')^2 + (BP')^2,$$
$$r^2 = x^2 + 2 a x \cos. (\beta - \alpha) + a^2 \cos.^2 (\beta - \alpha),$$
$$+ y^2 + 2 a y \sin. (\beta - \alpha) + a^2 \sin.^2 (\beta - \alpha),$$
$$= x^2 + y^2 + 2 a [x \cos. (\beta - \alpha) + y . \sin. (\beta - \alpha)] + a^2,$$
$$r = \sqrt{(x^2 + y^2 + a^2 + 2 a [x \cos. (\beta - \alpha) + y \sin. (\beta - \alpha)])} ; \quad (9)$$

$$\text{tang. } BAP' = \frac{BP'}{AP'},$$

$$\text{tang. } (\varphi - \alpha) = \frac{y + a \sin. (\beta - \alpha)}{x + a \cos. (\beta - \alpha)} ; \qquad (10)$$

and formulas (9) and (10) are the required formulas.

60. *Corollary.* If the origins are the same, we have

$$a = 0,$$

and the formulas (9) and (10) become

$$r = \sqrt{(x^2 + y^2)} ; \quad \text{tang. } (\varphi - \alpha) = \frac{y}{x}. \qquad (11)$$

61. Problem. *To transform from rectangular to polar coördinates.*

Solution. Let AX and AY (fig. 23.) be the rectangular axes. Let A_1 be the new origin, the coördinates of which are

$$AA' = a, \text{ and } A_1A' = b.$$

Let the inclination of the polar axis A_1C to the axis AX be α; so that, if A_1P' is drawn parallel to AX, we have

$$CA_1P' = \alpha.$$

The values of the rectangular coördinates

$$AP = x, \text{ and } BP = y,$$

are to be found in terms of the polar coördinates

$$A_1B = r, \text{ and } BA_1C = \varphi.$$

We have, then, in the right triangle BA_1P',

$$BA_1P' = BA_1C + CA_1P' = \varphi + \alpha,$$
$$BP' = A_1B . \sin. BA_1P' = r \sin. (\varphi + \alpha),$$
$$A_1P' = A_1B . \cos. BA_1P' = r \cos. (\varphi + \alpha);$$

whence

$$x = AA' + A'P = AA' + A_1P' = a + r \cos. (\varphi + \alpha) \quad (12)$$
$$y = PP' + P'B = A_1A' + P'B = b + r \sin. (\varphi + \alpha) \quad (13)$$

and (12) and (13) are the required values of x and y.

62. Corollary. If the origins are the same, we have

$$a = 0, \text{ and } b = 0,$$

and the formulas (12) and (13) become

$$x = r \cos. (\varphi + \alpha), \ y = r \sin. (\varphi + \alpha). \quad (14)$$

63. *Corollary.* If the origins are the same, and the polar axis coincides with the axis of x, we have

$$\alpha = 0,$$

and formula (14) becomes

$$x = r \cos. \varphi, \; y = r \sin. \varphi. \qquad (15)$$

64. *Problem.* To transform from one system of rectangular coördinates to another.

Solution. Let AX and AY (fig. 24.), be the axes of the first system ; and A_1X_1 and A_1Y_1 the new axes. Let the coördinates of the new origin A_1 be

$$AA' = a, \text{ and } A_1A' = b;$$

and let the inclination of the axis A_1X_1 to the axis AX be α, so that, if A_1R' is drawn parallel to AX, we have

$$X_1 A_1 R' = \alpha.$$

The values of the coördinates

$$AP = x, \text{ and } BP = y,$$

are to be found in terms of the coördinates

$$A_1P_1 = x_1, \text{ and } BP_1 = y_1.$$

Draw P_1R parallel to AX, and P_1R' parallel to AY. Since the sides of the angle B are respectively perpendicular to those of the angle P_1A_1R', they are equal, or

$$B = \alpha.$$

The right triangles A_1P_1R' and BP_1R give

$$P_1R' = A_1P_1 \sin. P_1A_1R' = x_1 \sin. \alpha,$$
$$A_1R' = A_1P_1 \cos. P_1A_1R' = x_1 \cos. \alpha;$$
$$P_1R = BP_1 \sin. B \qquad = y_1 \sin. \alpha,$$
$$BR = BP_1 \cos. B \qquad = y_1 \cos. \alpha.$$

4

we also have

$$A_1 P' = A_1 R' - P'R' = A_1 R' - P_1 R = x_1 \cos. \alpha - y_1 \sin. \alpha,$$
$$B P' = P R + B R = P_1 R' + B R = x_1 \sin. \alpha + y_1 \cos. \alpha;$$
$$A P = A A' + A' P = A A' + A_1 P',$$

or $x = \quad a + x_1 \cos. \alpha - y_1 \sin. \alpha;$ (16)

$$B P = P P' + B P' = A_1 A' + B P',$$

or $y = \quad b + x_1 \sin. \alpha + y_1 \cos. \alpha;$ (17)

and (16) and (17) are the required values of x and y.

65. *Corollary.* If the origins are the same, we have
$$a = 0, \text{ and } b = 0;$$
and the formulas (16) and (17) become
$$x = x_1 \cos. \alpha - y_1 \sin. \alpha, \tag{18}$$
$$y = x_1 \sin. \alpha + y_1 \cos. \alpha. \tag{19}$$

66. *Corollary.* If the directions of the axes are the same, we have
$$\alpha = 0, \ \sin. \alpha = 0, \ \cos. \alpha = 1 ;$$
and formulas (16) and (17) become
$$x = a + x_1, \ y = b + y_1. \tag{20}$$
If the new origin is, in this case, in the axis AX, we have
$$b = 0 ;$$
and formulas (20) become
$$x = a + x_1, \ y = y_1. \tag{21}$$
But if the new origin is in the axis AY, we have
$$a = 0,$$
and formulas (20) become
$$x = x_1, \ y = b + y_1. \tag{22}$$

Distance of two points.	Oblique axes.

67. Problem. *To express the distance between two points in terms of rectangular coördinates.*

Solution. The required result might be obtained by substituting in formula (5) the values of r, r', φ, and φ' obtained from formula (11), by taking x' and y' to correspond to r' and φ'. But it is more readily obtained by direct investigation.

Let the two points be B and R' (fig. 25.), whose coördinates are respectively x, y, and x', y'. Draw BR parallel to AX, and the right triangle $BB'R$ gives

$$BB'^2 = BR^2 + B'R^2 = (x' - x)^2 + (y' - y)^2$$
$$BB' = \sqrt{[(x' - x)^2 + (y' - y)^2]}. \qquad (23)$$

68. Corollary. If one of the points B is the origin, we have

$$x = 0, \; y = 0,$$

whence

$$AB' = \sqrt{(x'^2 + y'^2)}. \qquad (24)$$

69. Instead of the two axes being at right angles to each other, they are sometimes taken at any angle whatever; and instead of the distances from the axes, the lengths of lines drawn parallel to the axes are used.

In this case, the axes and coördinates are said to be *oblique*.

Thus if the axes are AX and AY (fig. 26.), the coördinates of B, B', &c., are, respectively,

$$x = AP = BR, \text{ and } y = BP = AR,$$

and

$$x' = AP' = B'R', \text{ and } y' = B'P' = AR', \text{ &c.}$$

70. Problem. *To transform from one system of oblique coördinates to another.*

Transformation of oblique coördinates.

Solution. Let AX and AY (fig. 27.) be the original axes, their mutual inclination being

$$A = \gamma.$$

Let the new axes be $A_1 X_1$ and $A_1 Y_1$, which are inclined to the axis AX by the angles α and β, so that, if $A_1 P'$ is drawn parallel to AX, we have

$$X_1 A_1 P' = \alpha, \text{ and } Y_1 A_1 P' = \beta.$$

Let the coördinates of A_1 be

$$AA' = a, \text{ and } A_1 A' = b.$$

The values of the coördinates

$$AP = x, \text{ and } BP = y,$$

are to be found in terms of

$$A_1 P_1 = x_1, \text{ and } BP_1 = y_1.$$

Draw $P_1 R$ parallel to AY, and $P_1 R'$ parallel to AX. We have, then,

$$P_1 R P' = P_1 R' P' = YAX = \gamma,$$

$$A_1 P_1 R = P_1 R P' - P_1 A_1 R = \gamma - \alpha;$$

$$BP_1 R' = Y_1 A_1 R = \beta,$$

$$B = P_1 R' P' - BP_1 R' = \gamma - \beta;$$

$$\sin. A_1 R P_1 : \sin. A_1 P_1 R = A_1 P_1 : A_1 R,$$

or $\quad\quad \sin. \gamma : \sin. (\gamma - \alpha) = x_1 : A_1 R,$

and $\quad\quad A_1 R = \dfrac{x_1 \sin. (\gamma - \alpha)}{\sin. \gamma};$

$$\sin. A_1 R P_1 : \sin. P_1 A_1 R = A_1 P_1 : P_1 R,$$

or $\quad\quad \sin. \gamma : \sin. \alpha = x_1 : P_1 R,$

and $\quad\quad P_1 R = P' R' = \dfrac{x_1 \sin. \alpha}{\sin. \gamma};$

Transformation of oblique coördinates.

$$\text{sin. } BR'P_1 : \text{sin. } B = BP_1 : P_1R',$$

or $\qquad \text{sin. } \gamma : \text{sin. } (\gamma - \beta) = y_1 : P_1R',$

and $\qquad P_1R' = RP' = \dfrac{y_1 \text{ sin. } (\gamma - \beta)}{\text{sin. } \gamma};$

$$\text{sin. } BR'P_1 : \text{sin. } BP_1R' = BP_1 : BR',$$

or $\qquad \text{sin. } \gamma : \text{sin. } \beta = y_1 : BR',$

and $\qquad BR' = \dfrac{y_1 \text{ sin. } \beta}{\text{sin. } \gamma};$

$$AP = AA' + A'P = AA' + A_1P' = AA' + A_1R + RP',$$

or $\qquad x = a + \dfrac{x_1 \text{ sin. } (\gamma - \alpha) + y_1 \text{ sin. } (\gamma - \beta)}{\text{sin. } \gamma};$ $\qquad$ (25)

$$BP = PP' + P'B = A_1A' + P'R' + R'B,$$

or $\qquad y = b + \dfrac{x_1 \text{ sin. } \alpha + y_1 \text{ sin. } \beta}{\text{sin. } \gamma};$ $\qquad$ (26)

and (25) and (26) are the required values of x and y.

71. *Corollary.* If the original axes are rectangular, we have

$$\gamma = 90°, \quad \text{sin. } \gamma = 1,$$

and formulas (25) and (26) become

$$x = a + x_1 \text{ cos. } \alpha + y_1 \text{ cos. } \beta, \qquad (27)$$
$$y = b + x_1 \text{ sin. } \alpha + y_1 \text{ sin. } \beta. \qquad (28)$$

72. *Corollary.* If the new axes are rectangular, we have

$$\beta = 90° + \alpha, \quad \text{sin. } \beta = \text{cos. } \alpha,$$
$$\text{sin. } (\gamma - \beta) = \text{sin. } (\gamma - \alpha - 90°) = -\text{cos. } (\gamma - \alpha);$$

and formulas (25) and (26) become

$$x = a + \dfrac{x_1 \text{ sin. } (\gamma - \alpha) - y_1 \text{ cos. } (\gamma - \alpha)}{\text{sin. } \gamma}, \qquad (29)$$

$$y = b + \dfrac{x_1 \text{ sin. } \alpha + y_1 \text{ cos. } \alpha}{\text{sin. } \gamma}. \qquad (30)$$

4*

73. Problem. *To determine the position of points in space.*

Solution. The most natural method of determining the position of a point in space is to determine the position of some plane passing through the point, and then to determine the position of the point in this plane. For this purpose some fixed axis AC (fig. 28.) is assumed, and some fixed plane CAD passing through this axis. The plane CAE, which passes through the point B and the axis AC, is determined by the angle EAD, which it makes with the fixed plane CAD. The position of the point B in the plane CAB is determined by the radius vector AB and the angle BAC, which this radius makes with the axis. The same method may be adopted for any other points B', B'', &c., which are not given on the figure, as they would only render it confused. We may, then, denote these radii vectores of the points B, B', &c. by r, r', &c.; the angles which these radii make with the axis AC by φ, φ', &c.; and the angles which the planes in which they are contained make with the fixed plane by θ, θ', &c.

74. A system of *rectangular coördinates* in space has been adopted similar to those in plane, and possessing the same practical advantages of simplicity.

For this purpose three planes XAY, YAZ, and XAZ (fig. 29.) are drawn perpendicular to each other, and the rectangular coördinates of a point are its distances from these planes. Thus, if the point B is taken, and the perpendiculars BP, BQ, and BR are drawn perpendicular to the given planes, these distances are the rectangular coördinates of B. If these coördinates are given, the point B is determined, by taking

$$AL = BR, \quad AM = BQ, \text{ and } AN = BP,$$

Projection of a point.

and drawing planes through the points L, M, and N parallel respectively to the given planes, and their intersection B is the required point.

The intersections AX, AY, and AZ of the three given planes are called the *axes*.

If the coördinates of the point B are

$$x = AL = BR = NQ = MP,$$

$$y = AM = BQ = LP = NR,$$

$$z = AN = BP = MR = LQ,$$

the axis AX is called the axis of x, AY is called the axis of y, and AZ the axis of z; the plane XAY is called the plane of xy, the plane XAZ is called the plane xz, and the plane YAZ is called the plane of yz. The coördinates of B', B'', &c. are, in the same way,

$$x' = AL', \ y' = B'Q', \ z' = L'Q',$$

and

$$x'' = AL'', \ y'' = B''Q'', \ z'' = R''Q'', \ \&c. \ \&c.$$

75. The foot of the perpendicular let fall from a point upon a plane is called *the projection of the point upon the plane.*

Thus the projection of B upon the plane of xy is P, the coördinates of which are x and y; the projection upon the plane of xz is Q, the coördinates of which are x and z; the projection upon the plane of yz is R, the coördinates of which are y and z.

76. *Problem. To transform from rectangular co-ordinates to the polar coördinates of art. 73, the origins being the same, the polar axis being the axis of x, and the fixed plane the plane of xy.*

Solution. Let AX, AY, AZ (fig. 30.) be the rectangular axes. Let XAD be a plane passing through the point B.

The values of

$$AL = x, \quad BQ = y, \text{ and } CQ = z,$$

are to be found in terms of

$$AB = r, \quad BAX = \varphi, \text{ and } DAY = \theta.$$

Join BL. In the right triangles ABL or BLQ, we have the angle

$$LBQ = \theta,$$

because the sides LB and BQ are parallel to AD and AY; we also have

$$x = AL = AB \cos. BAL = r \cos. \varphi, \tag{31}$$
$$BL = AB \sin. BAL = r \sin. \varphi\ ;$$
$$y = BQ = BL \cos. LBQ = r \sin. \varphi \cos. \theta, \tag{32}$$
$$z = LQ = BL \sin. LBQ = r \sin. \varphi \sin. \theta. \tag{33}$$

77. *Problem.* *To transform from the polar coördinates of art.* 73, *to rectangular coördinates, the origins being the same, the polar axis being the axis of x, and the fixed plane the plane of x y.*

Using the figure and notation of· the preceding article, the values of r, φ, and θ, are to be found in terms of x, y, and z. They may be immediately found from equations (31), (32), and (33). The sum of the squares of these equations is

$$x^2 + y^2 + z^2 = r^2 (\cos.^2 \varphi + \sin.^2 \varphi \cos.^2 \theta + \sin.^2 \varphi \sin.^2 \theta)$$
$$= r^2 [\cos.^2 \varphi + \sin.^2 \varphi (\cos.^2 \theta + \sin.^2 \theta)]$$
$$= r^2 (\cos.^2 \varphi + \sin.^2 \varphi) = r^2\ ;$$

because,

$$1 = \cos^2 \theta + \sin.^2 \theta = \cos.^2 \varphi + \sin.^2 \varphi.$$

Distance of two points in space.

Hence

$$r = \sqrt{(x^2 + y^2 + z^2)}. \qquad (34)$$

and (31) gives

$$\cos. \varphi = \frac{x}{r} = \frac{x}{\sqrt{(x^2 + y^2 + z^2)}}. \qquad (35)$$

The quotient of (33) divided by (32) is

$$\frac{\sin. \vartheta}{\cos. \vartheta} = \tan. \vartheta = \frac{z}{y}. \qquad (36)$$

78. *Problem.* *To find the distance apart of two points.*

Solution. Let B, B' (fig. 31.) be the points, and P, P' their projections upon the plane of xy. Join PP', and draw BE parallel to PP'. Since P, P' are two points in the plane YAX, we have, by equation (23),

$$PP'^2 = (x' - x)^2 + (y' - y)^2;$$

and in the right triangle BEB',

$$B'E = B'P' - BP = z' - z,$$
$$BB'^2 = BE^2 + B'E^2 = PP'^2 + B'E^2,$$
$$= (x' - x)^2 + (y' - y)^2 + (z' - z)^2;$$
$$BB' = \sqrt{[(x' - x)^2 + (y' - y)^2 + (z' - z)^2]}. \qquad (37)$$

79. *Corollary.* If one of the points, as B', is the origin, we have

$$x' = 0,\ y' = 0,\ z' = 0,$$

and formula (37) becomes

$$AB = \sqrt{(x^2 + y^2 + z^2)}, \qquad (38)$$

which agrees with equation (34).

80. The line PP', which joins the projections of the two extremities B, B' of the line BB', is called the *projection of the line BB' upon the plane of x y*.

81. *Corollary.* If the angle $B'BE$, which is the inclination of the line BB' to its projection or to the plane of $y x$, is denoted by λ, the right triangle $B'BE$ gives

$$B'E = BB' \cos. B'BE,$$

or

$$PP' = BB' \cos. \lambda;$$

that is, the projection of a straight line upon a plane is equal to the product of the line multiplied by the cosine of its inclination to the plane.

82. If planes BPL, $B'P'L'$, are drawn through the extremities B, B' of a line BB', perpendicular to an axis AX, the part LL' intercepted between these lines is called the *projection of the line upon this axis*.

83. *Corollary.* Since

$$LL' = AL' - AL = x' - x,$$

the projection of a line upon an axis is equal to the difference of the corresponding ordinates of its extremities.

The projection of the radius vector AB is AL, or the corresponding coördinate of its extremity.

84. *Corollary.* It follows from equation (37), that the square of a line is equal to the sum of the squares of its projections upon the three rectangular axes.

Sum of the squares of the angles made by a line with the axes.

85. *Corollary.* If the inclination of the line $\stackrel{.}{AB}$ to the axis AX is denoted by φ, we have, by drawing LS parallel to AB to meet the plane $B'P'L'$ in S,

$$LS = BB', \ SLL' = \varphi,$$

$$LL' = LS \cos. \ SLL',$$

$$LL' = BB' \cos. \ \varphi \ ;$$

that is, the projection of a straight line upon an axis is equal to the product of the line multiplied by the cosine of its inclination to the axis.

86. *Corollary.* If ψ is the inclination of the line to the axis of y, and ω its inclination to the axis of z, its projections upon the axes are, respectively,

$$BB' \cos. \ \psi, \text{ and } BB' \cos. \ \omega \ ;$$

so that, by art. 82,

$$BB'^2 = BB'^2 \cos.^2 \varphi + BB'^2 \cos.^2 \psi + BB'^2 \cos.^2 \omega,$$

or, dividing by BB'^2,

$$1 = \cos.^2 \varphi + \cos.^2 \psi + \cos^2 \omega \ ;$$

that is, *the sum of the squares of the angles of the angles which a line makes with three rectangular axes is equal to unity.*

87. A different system of polar coördinates from that of art. 73, is often used upon account of its symmetrical character. It consists in determining the direction of the radius vector by the angle which it makes with three rectangular axes.

88. *Corollary.* If the angles φ, ψ, and ω denote the angles which the radius vector makes with the axes of x, y, and z, we have, by arts. 83 and 84, $\;\;s\;\;\;\cdots\;\;\;\;\;\;s\;\cdot$

$$x = r \cos. \varphi, \quad y = r \cos. \psi, \quad z = r \cos. \omega; \qquad (39)$$

$$\cos.^2 \varphi + \cos.^2 \psi + \cos.^2 \omega = 1,$$

which will serve to transform from rectangular to polar coördinates in the system of the preceding article.

89. *Oblique coördinates* are sometimes used similar to oblique coördinates in a plane; thus, if the axes AX, AY, and AZ (fig. 29.) had been oblique to each other, and the other lines drawn parallel to the axes, the point B would be determined by the oblique coördinates

$$AL = x, \quad LQ = y, \quad QB = z;$$

and, in the same way, for other points B', B'', &c.

90. *Problem.* *To transform from rectangular to oblique coördinates.*

Let AX, AY, AZ, (fig. 32.) be the rectangular axes, and $A_1 X_1$, $A_1 Y_1$, $A_1 Z_1$, the oblique axes. Let the coördinates of the new origin be

$$AA' = a, \; A_1 A'' = b, \; A'A'' = c;$$

and let the inclination to the axes AX, AY, and AZ of those axes $A_1 X_1$ be, respectively, α, α', α''; let those of the axis $A_1 Y_1$ be β, β', β''; and those of the axis $A_1 Z_1$ be γ, γ', γ''; these angles must be subject to the condition of art. 84; that is,

$$\cos.^2 \alpha + \cos.^2 \alpha' + \cos.^2 \alpha'' = 1,$$

$$\cos.^2 \beta + \cos.^2 \beta' + \cos.^2 \beta'' = 1,$$

$$\cos.^2 \gamma + \cos.^2 \gamma' + \cos.^2 \gamma'' = 1.$$

Oblique transformed to rectangular coördinates.

The values of the rectangular coördinates

$$AL = x, \quad BQ = y, \quad LQ = z,$$

are to be found in terms of

$$A_1 L_1 = x_1, \quad BQ_1 = y_1, \quad L_1 Q_1 = z_1.$$

Let L' and Q' be the projections of the points L_1 and Q_1 upon the axis of x. Since $A'L'$, $L'Q'$, and $Q'L$ are the projections, respectively, of $A_1 L_1$, $L_1 Q_1$, and $Q_1 B$, upon the axis of x; and A, L_1, $Q_1 B$, and $L_1 Q_1$, are respectively parallel to the axes of x_1, y_1, and z_1, and, therefore, inclined to the axis of x by the angles α, β, and γ, we have

$$A'L' = A_1 L_1 \; \cos. \; \alpha = x_1 \; \cos. \; \alpha,$$

$$QL = Q_1 B \; \cos. \; \beta = y_1 \; \cos. \; \beta,$$

$$L'Q' = L_1 Q_1 \; \cos. \; \gamma = z_1 \; \cos. \; \gamma;$$

so that

$$AL = AA' + A'L' + Q'L + L'Q'$$

gives

$$x = a + x_1 \; \cos. \; \alpha + y_1 \; \cos. \; \beta + z_1 \; \cos. \; \gamma. \quad (40)$$

In the same way we might find

$$y = b + x_1 \; \cos. \; \alpha' + y_1 \; \cos. \; \beta' + z_1 \; \cos. \; \gamma'. \quad (41)$$

$$z = c + x_1 \; \cos. \; \alpha'' + y_1 \; \cos. \; \beta'' + z_1 \; \cos. \; \gamma''; \quad (42)$$

so that equations (40), (41), and (42) are the required equations.

91. *Corollary.* If the new axes are also rectangular, equations (40), (41), and (42) may still be used, but the angles α, β, γ, α', β', γ', α'', β'', and γ'' will be subject to certain conditions, which are thus obtained. Let r be the radius vector drawn from A_1 to B, and let the angles which r makes with

Angle of two lines.

the axes AX, AY, AZ, A_1X_1, A_1Y_1, A_1Z_1, be respectively φ, ψ, ω, φ_1, ψ_1, ω_1, we shall have, by art. 83,

$$x = AL = AA' + A'L = a + r \cos. \varphi,$$

$$x_1 = r \cos. \varphi_1, \ y_1 = r \cos. \psi_1, \ z_1 = r \cos. \omega_1;$$

which may be substituted in equation (40). If, in the result, we suppress the common term a, and the common factor r, we have

$$\cos. \varphi = \cos. \alpha \cos. \varphi_1 + \cos. \beta \cos. \psi_1 + \cos. \gamma \cos. \omega_1; \quad (43)$$

which expresses the angle φ made by two lines, one of which is inclined to the three axes of x_1, y_1, z_1, by the angles α, β, γ; and the other by the angles φ_1, ψ_1, ω_1.

This formula may then be used for determining the angle which any two lines make with each other, and which are inclined to the axes of x, y, z by given angles; to determining the angles, for instance, which the axes of x_1, y_1, z_1 make with each other. But these axes are perpendicular to each other, and therefore we have for the angles of x_1 and y_1, of x_1 and z_1, of y_1 and z_1, respectively,

$$\cos. 90° = 0 = \cos. \alpha \cos. \beta + \cos. \alpha' \cos. \beta' + \cos. \alpha'' \cos. \beta'', \quad (44)$$

$$\cos. 90° = 0 = \cos. \alpha \cos. \gamma + \cos. \alpha' \cos. \gamma' + \cos. \alpha'' \cos. \gamma'', \quad (45)$$

$$\cos. 90° = 0 = \cos. \beta \cos. \gamma + \cos. \beta' \cos. \gamma' + \cos. \beta'' \cos. \gamma''. \quad (46)$$

92. *Corollary.* By applying the preceding formulas to the axes of x, y, z, referred to those of x_1, y_1, z_1, we have

$$\cos.^2 \alpha + \cos.^2 \beta + \cos.^2 \gamma = 1, \quad (47)$$

$$\cos.^2 \alpha' + \cos.^2 \beta' + \cos.^2 \gamma' = 1, \quad (48)$$

$$\cos.^2 \alpha'' + \cos.^2 \beta'' + \cos.^2 \gamma'' = 1; \quad (49)$$

Change of origin.

$$\cos. \alpha \cos. \alpha' + \cos. \beta \cos. \beta' + \cos. \gamma \cos. \gamma' = 0, \qquad (50)$$

$$\cos \alpha \cos. \alpha'' + \cos. \beta \cos. \beta'' + \cos. \gamma \cos. \gamma'' = 0, \qquad (51)$$

$$\cos. \alpha' \cos. \alpha'' + \cos. \beta' \cos. \beta'' + \cos. \gamma' \cos. \gamma'' = 0. \qquad (52)$$

93. *Corollary.* If the origin is changed, but not the directions of the axes, we have

$$\alpha = 0, \quad \beta = 90°, \ \gamma = 90°,$$

$$\alpha' = 90°, \ \beta' = 0, \quad \gamma' = 90°,$$

$$\alpha'' = 90°, \ \beta'' = 90°, \ \gamma'' = 0;$$

and equations (40, 41, 42) become

$$x = a + x_1 \qquad (53)$$

$$y = b + y_1 \qquad (54)$$

$$z = c + z_1. \qquad (55)$$

CHAPTER IV.

EQUATIONS OF LOCI.

94. When a geometrical question regarding position leads to a number of equations less than that of the unknown quantities, it is *indeterminate*, and usually admits of many solutions; that is, there are usually a series of points which solve it, and this series of points is called *the locus of the question, or of the equations to which it leads.*

95. *The equation of the locus of a geometrical question is found by referring the positions of its points to coördinates, as in the preceding chapter, and expressing algebraically the conditions of the question.*

96. *Scholium.* Instead of denoting angles by degrees, minutes, &c.; we shall hereafter denote them by the lengths of the arcs which measure them upon the circumference of a circle whose radius is unity, and shall denote by π the semicircumference of this circle, which is nearly 3·1415926.

The angle of 90°, or the right angle, is thus denoted by $\frac{1}{2}\pi$, the angle of 180°, or two right angles, by π, and the angle of 360°, or four right angles, by 2π.

97. *Corollary.* The arc which measures an angle θ in the

Circle.
---

circle whose radius is R is $R\vartheta$, because similar arcs are proportional to their radii, and ϑ is the length if the radius is unity.

98. EXAMPLES.

1. Find the equation of the locus of all the points in a plane, which are at a given distance from a given point in that plane. This locus is *the circumference of the circle*.

Solution. Let the given point A (fig. 33.) be assumed as the origin of coördinates, and let $R =$ the given distance.

If the polar coördinates of art. 44 are used, we have for each of the required points, as M,

$$r = R ; \qquad (56)$$

so that this equation is that of the required locus.

Corollary 1. *Equation* (56) *is the polar equation of the circle whose radius is R, and centre at the origin.*

Corollary 2. Equation (56) may be referred to other polar axes by arts. 48 and 49. Thus for the point A_1, for instance, for which

$$AA_1 = a = - R$$

equation (3) becomes

$$r = \sqrt{(R^2 + r_1^2 - 2 R r_1 \cos. \varphi_1)}$$

which substituted in (56) gives, by squaring and reducing,

$$r_1^2 - 2 R r_1 \cos. \varphi_1 = 0 ;$$

or we may divide by r_1, since r_1 is not generally equal to zero, and the equation is

$$r_1 = 2 R \cos. \varphi_1 \qquad (57)$$

5*

which is the polar equation of a circle whose radius is R, the origin being upon the circumference, and the line drawn to the centre being the axis.

Corollary 3. Equation (56) may, by art. 60, be referred to rectangular coördinates; and equations (11) being substituted in (56), and the result being squared, we have

$$x^2 + y^2 = R^2 \qquad (58)$$

which is the equation of a circle whose radius is R, referred to rectangular coördinates, the origin of which is the centre.

Corollary 4. Equation (58) may, by art. 66, be referred to any rectangular coördinates. Thus the axes $A_2 X_2$, $A_2 Y_2$, for which the coördinates of A are $A_2 A'$ and $A A'$, so that

$$a = - A_2 A' = - a',$$
$$b = - A A' = - b',$$

give

$$x = x_2 - a', \ y = y_2 - b',$$

.which, substituted in (58), give

$$(x_2 - a')^2 + (y_2 - b')^2 = R^2, \qquad (59)$$

which is the equation of a circle, referred to rectangular coördinates, the radius of the circle being R, and the coördinates of the centre a' and b'.

Corollary 5. For the point A_1 we have

$$a' = R, \ b' = 0,$$

so that for this point (59) becomes

$$(x_1 - R)^2 + y_1^2 = R^2,$$

or

$$x_1^2 - 2 R x_1 + R^2 + y_1^2 = R^2$$

$$y_1^2 = 2 R x_1 - x_1^2, \qquad (60)$$

which is the equation of a circle, whose radius is R, referred to rectangular coördinates, the origin of which is upon the circumference, and the axis of x_1 is the diameter.

2. Find the equation of the locus of all the points in space, which are at a given distance from a given point. This locus is *the surface of the sphere.*

Solution. Let the given point be assumed as the origin of coördinates, and let

$$R = \text{the given distance.}$$

If polar coördinates are used, we have for each of the required points

$$r = R; \qquad (61)$$

which is, therefore, the polar equation of a sphere, whose radius is R, and centre at the origin.

Corollary 1. Equation (61) may be referred to rectangular coördinates, by art. 77; and if equation (34) is substituted in (61), and the result squared, we have

$$x^2 + y^2 + z^2 = R^2; \qquad (62)$$

which is the equation of the sphere whose radius is R, referred to rectangular coördinates, the origin of which is the centre.

Corollary 2. Equation (62) may, by art. 93, be referred to any rectangular coördinates, and the substitution of equations (53, 54, 55) in (62) gives

$$(x_1 + a)^2 + (y_1 + b)^2 + (z_1 + c)^2 = R^2, \qquad (63)$$

which is the equation of a sphere referred to rectangular

coördinates, the radius of the sphere being R, and the coördinates of the centre — *a,* — *b, and* — *c.*

3. Find the equation of the locus of all the points in a plane, of which the sum of the distances of each point from two given points in that plane is equal to a given line. This locus is called *the ellipse,* and the given points are called *its foci.*

Solution. Let F and F' (fig. 34.) be the foci, let F be the polar origin, let the line FF' joining the foci be the polar axis, and let

$$2c = FF' = \text{distance between the foci,}$$

$$2A = \text{the given length;}$$

where the length A is not to be confounded with the point A of the figure.

If, then, we put in equation (6)

$$r' = FF' = 2c,$$

we have for the distance MF' of each point M from F',

$$MF' = \sqrt{(r^2 + 4c^2 - 4cr \cos. \varphi)};$$

so that

$$FM + MF' = 2A = r + \sqrt{(r^2 + 4c^2 - 4cr \cos. \varphi)}$$

$$\sqrt{(r^2 + 4c^2 - 4cr \cos. \varphi)} = 2A - r,$$

and squaring and reducing

$$4c^2 - 4cr \cos. \varphi = 4A^2 - 4Ar$$

$$(A - c \cos. \varphi) r = A^2 - c^2$$

$$r = \frac{A^2 - c^2}{A - c \cos. \varphi}, \qquad (64)$$

which is the polar equation of the ellipse, one of the foci being the origin, and the axis being the line joining the foci, which is called the transverse axis, if it is produced to meet the curve.

Corollary 1. For the point C where the ellipse cuts the axis, we have

$$\varphi = 0, \ \cos. \ \varphi = 1,$$

$$FC = r = \frac{A^2 - c^2}{A - c} = A + c.$$

Corollary 2. For the point C', where the ellipse cuts the axis produced, we have

$$\varphi = \pi, \ \cos. \ \varphi = -1,$$

$$FC' = r = \frac{A^2 - c^2}{A + c} = A - c.$$

$$CC' = FC + FC' = 2\,A;$$

so that *the transverse axis is equal to the sum of the distances of each point of the ellipse from the two foci.*

Corollary 3. If FF' is bisected at A, we have

$$AF = AF' = c,$$

$$AC = FC - AF = A + c - c = A = \tfrac{1}{2}\,CC' = AC';$$

A is called *the centre of the ellipse.*

Corollary 4. If we put

$$B^2 = A^2 - c^2,$$

(64) becomes

$$r = \frac{B^2}{A - c \cos. \varphi}. \tag{65}$$

Corollary 5. If we put

$$e = \frac{c}{A} = \frac{2\,c}{2\,A};$$

e is called the eccentricity of the ellipse, and is the ratio of the distance between the foci divided by the transverse axis.

Hence $c = A\,e,$

and this, substituted in (64), gives

$$r = \frac{A^2\,(1 - e^2)}{A\,(1 - e \cos. \varphi)} = \frac{A\,(1 - e^2)}{1 - e \cos. \varphi}. \tag{66}$$

If we also put

$$FC' = P = A - c = A\,(1 - e),$$

(66) may be put in the form

$$r = \frac{A\,(1 - e)\,(1 + e)}{1 - e \cos. \varphi} = \frac{P\,(1 + e)}{1 - e \cos. \varphi}. \tag{67}$$

Corollary 6. The equation of the ellipse may be referred to rectangular coördinates by arts. 59 and 60. Thus, if we take the point A for the origin, and AC for the axis of x, we have

$$a = FA = c,$$

$$\alpha = \beta = 0, \text{ sin. } (\beta - \alpha) = 0, \text{ cos. } (\beta - \alpha) = 1.$$

whence (9) and (10) become

$$r = \sqrt{(x^2 + y^2 + c^2 + 2\,c\,x)},$$

$$\text{tang. } \varphi = \frac{y}{x + c};$$

Ellipse.	Conjugate axis.

and

$$\cos. \varphi = \frac{1}{\sec. \varphi} = \frac{1}{\sqrt{(1+\tan.^2\varphi)}} = \frac{x+c}{\sqrt{(y^2+2cx+c^2+x^2)}}$$

$$\cos. \varphi = \frac{x+c}{r};$$

$$r \cos. \varphi = x + c.$$

Now equation (64) freed from fractions is

$$A r - c r \cos. \varphi = A^2 - c^2;$$

in which, if we substitute the preceding values, we have

$$A\sqrt{(x^2 + y^2 + c^2 + 2cx)} - cx - c^2 = A^2 - c^2,$$

or

$$A\sqrt{(x^2 + y^2 + c^2 + 2cx)} = A^2 + cx.$$

The square of which is, when reduced,

$$A^2 x^2 + A^2 y^2 + A^2 c^2 = A^4 + c^2 x^2,$$

or

$$(A^2 - c^2) x^2 + A^2 y^2 = A^4 - A^2 c^2 = A^2 (A^2 - c^2),$$

or substituting B^2

$$B^2 x^2 + A^2 y^2 = A^2 B^2; \qquad (68)$$

which, divided by $A^2 B^2$, is

$$\frac{x^2}{A^2} + \frac{y^2}{B^2} = 1. \qquad (69)$$

Corollary 7. The part BB' of the axis of y included within the ellipse is called *its conjugate axis.*

We have for the points B and B', art. 58,

$$x = 0, \; y' = AB \text{ or } = -AB';$$

Ellipse.	Conjugate axes.

which, substituted in (68), gives

$$A^2 y'^2 = A^2 B^2,$$
$$y' = \pm B = + B \text{ or } = - B;$$

so that $\qquad AB = AB' = B,$

and B *is equal to the semiconjugate axis.*

Corollary 8. Equations (68) and (69) are the *equations of an ellipse referred to rectangular coördinates, the centre of the ellipse being the origin, the transverse axis* $2A$ *being the axis of* x, *and the conjugate axis* $2B$ *being the axis of* y.

Corollary 9. The equation of the ellipse may, by art. 71, be referred to oblique axes. Thus, if the origin is unchanged, we have

$$a = 0, \quad b = 0,$$

and equations (27) and (28) become

$$x = x_1 \cos. \alpha + y_1 \cos. \beta,$$
$$y = x_1 \sin. \alpha + y_1 \sin. \beta;$$

which, substituted in (68), give, by simple reduction,

$$(B^2 \cos.^2\alpha + A^2 \sin.^2\alpha)x_1^2 + 2(B^2 \cos.\alpha \cos.\beta + A^2 \sin.\alpha \sin.\beta) x_1 y_1$$
$$+ (B^2 \cos.^2\beta + A^2 \sin.^2\beta) y_1^2 = A^2 B^2 \qquad (70)$$

Corollary 10. If α and β, instead of being taken arbitrarily, are so taken that we have

$$B^2 \cos. \alpha \cos. \beta + A^2 \sin. \alpha \sin. \beta = 0, \qquad (71)$$

or $\qquad A^2 \sin. \alpha \sin. \beta = - B^2 \cos. \alpha \cos. \beta,$

or, dividing by $A^2 \cos. \alpha \cos. \beta,$

$$\text{tang. } \alpha \text{ tang. } \beta = - \frac{B^2}{A^2}, \qquad (72)$$

the axes are said to be *conjugate* to each other.

Equation (70) is then reduced to

$$(B^2\cos.^{2}\alpha + A^2\sin.^{2}\alpha)x_1^2 + (B^2\cos.^{2}\beta + A^2\sin.^{2}\beta)y_1^2 = A^2B^2. \quad (73)$$

Corollary 11. If $C_1 AC'_1$ (fig. 35.) is the axis of x_1, and $B_1 A B'_1$ is the axis of y_1, we have for the points C_1 and C'_1, where the axis of x_1 meets the curve

$$y_1 = 0, \; x'_1 = AC_1 \text{ or } - AC'_1 \;;$$

which, substituted in (73), give

$$(B^2 \cos.^2 \alpha + A^2 \sin.^2 \alpha) \, x'^2_1 = A^2B^2$$

$$x'_1 = \pm \frac{AB}{\sqrt{(B^2 \cos.^2 \alpha + A^2 \sin.^2 \alpha)}} \;;$$

so that the distances AC_1 and AC'_1 are equal, and in general any line, which passes through the centre and terminates in the curve, is bisected, and is hence called a *diameter*. The axes of x_1 and y_1, which are subject to the condition of equation (71) or 72), are called *conjugate diameters, and equation (73) is the equation of the ellipse referred to conjugate diameters, which are inclined by the angles α and β to the transverse axis.*

Corollary 12. If we take

$$A' = AC_1 = AC'_1,$$
$$B' = AB_1 = AB'_1 ;$$

we have

$$A' = \frac{AB}{\sqrt{(B^2 \cos.^2 \alpha + A^2 \sin.^2 \alpha)}},$$

$$B' = \frac{AB}{\sqrt{(B^2 \cos.^2 \beta + A^2 \sin.^2 \beta)}} ;$$

6

so that

$$\sqrt{(B^2 \cos^2 \alpha + A^2 \sin^2 \alpha)} = \frac{AB}{A'},$$

$$\sqrt{(B^2 \cos^2 \beta + A^2 \sin^2 \beta)} = \frac{AB}{B'};$$

which, substituted in (73), and the result divided by A^2B^2, give

$$\frac{x_1^2}{A'^2} + \frac{y_1^2}{B'^2} = 1, \tag{74}$$

or

$$B'^2 x_1^2 + A'^2 y_1^2 = A'^2 B'^2; \tag{75}$$

which are, therefore, precisely similar in form to equations (68) and (69); *and they are the equations of the ellipse referred to the conjugate diameters* 2 A' *and* 2 B'.

4. Find the equation of the locus of all the points in a plane, of which the difference of the distances of each point from two given points in that plane is equal to a given line. This locus is called *the hyperbola,* and the given points are called *its foci.*

Solution. Let F and F' (fig. 36.) be the foci, let F be the polar origin, let the line FF' joining the foci be the polar axis, and let

$$2 c = FF' = \text{the distance between the foci,}$$
$$2 A = \text{the given length.}$$

If we put, in equation 6,

$$r' = FF' = 2 c,$$

we have, for the distance of each point M from F',

$$MF' = \sqrt{(r^2 + 4 c^2 - 4 c r \cos \varphi)};$$

so that

$$MF - MF' = 2\,A = r - \sqrt{(r^2 + 4\,c^2 - 4\,c\,r\cos.\,\varphi)},$$

$$\sqrt{(r^2 + 4\,c^2 - 4\,c\,r\cos.\,\varphi)} = r - 2\,A.$$

Squaring and reducing, we have

$$4\,c^2 - 4\,c\,r\cos.\,\varphi = -4\,A\,r + 4\,A^2$$

$$(A - c\cos.\,\varphi)\,r = A^2 - c^2$$

$$r = \frac{A^2 - c^2}{A - c\cos.\,\varphi} = \frac{c^2 - A^2}{c\cos.\,\varphi - A};\qquad (76)$$

which is the polar equation of the hyperbola, one of the foci being the origin, and the axis being the line which joins the foci, the part of which CC', intersected by the curve, is called the transverse axis.

Corollary 1. If equation (76) is compared with (64), it appears that these equations have the same form, and only differ in the circumstance, that c is less than A for the ellipse, and greater than A for the hyperbola. In the ellipse, then, the value of r is always positive, because the numerator $A^2 - c^2$ is positive, and so is the denominator $A - c\cos.\,\varphi$. For $c\cos.\,\varphi$ is less than c, and therefore less than A. But in the hyperbola, while the numerator $c^2 - A^2$ is positive, the denominator $c\cos.\,\varphi - A$ is only positive when

$$c\cos.\,\varphi > A,$$

or

$$\cos.\,\varphi > \frac{A}{c}.$$

If then we take $\cos.\,\varphi_0 = \dfrac{A}{c}$, φ must be confined to the limits φ_0 and $-\varphi_0$.

Corollary 2. The above solution is limited to the hypothesis, that FM is greater than $F'M$, but if it were supposed less, we should have the equation of another curve situated with reference to the foci F' and F precisely as the curve of equation (76) is with regard to the foci F and F'.

Since both these curves satisfy the conditions of the problem, they are included in the common name of *an hyperbola*, and are called *its branches*.

To find the equation of the second branch referred to the same polar coördinates as those already used, we have

$$F'M_1 - FM_1 = 2A = \sqrt{(r^2 + 4c^2 - 4cr\cos.\varphi)} - r,$$

$$\sqrt{(r^2 + 4c^2 - 4cr\cos.\varphi)} = 2A + r.$$

Squaring and reducing, we have

$$c^2 - cr\cos.\varphi = A^2 + Ar,$$

$$(A + c\cos.\varphi)r = c^2 - A^2,$$

$$r = \frac{c^2 - A^2}{A + c\cos.\varphi}; \qquad (77)$$

which is the polar equation of the branch $C'M_1$, the focus F being the origin, and the line joining the foci being the axis.

Corollary 3. The numerator of equation (77) is positive, but the denominator is negative, when

$$A + c\cos.\varphi < 0,$$

or $$\cos.\varphi < -\frac{c}{A},$$

or $$\cos.\varphi < -\cos.\varphi_0;$$

Transverse axis of hyperbola.	Centre.

or when φ is included between the limits $\pi - \varphi_0$ and $\pi + \varphi_0$, so that φ must not be taken between these limits.

Corollary 4. For the point C, where the first branch of the hyperbola cuts the axis, we have, by equation (76),

$$\varphi = 0, \quad \cos. \varphi = 1,$$

$$FC = r = \frac{c^2 - A^2}{c - A} = c + A.$$

Corollary 5. For the point C', where the second branch cuts the axis, we have, by (77),

$$\varphi = 0, \quad \cos. \varphi = 1,$$

$$FC' = r = \frac{c^2 - A^2}{c + A} = c - A.$$

Hence　　　　$CC' = FC - FC' = 2\,A \; ;$

and the transverse axis is equal to the difference of the distances of each point of the hyperbola from the two foci.

Corollary 6. If FF' is bisected at A, we have

$$AF = AF' = c,$$

$$AC = AF - AC' = A + c - c = A = \tfrac{1}{2}CC = AC.$$

A is called *the centre of the hyperbola.*

Corollary 7. If we put

$$B^2 = c^2 - A^2,$$

6*

(76) and (77) become

$$r = \frac{B^2}{c \cos. \varphi - A} \tag{78}$$

$$r = \frac{B^2}{A + c \cos. \varphi}. \tag{79}$$

If this value of B^2 is compared with that of the ellipse of corollary 4, we see that it is, in form, the negative of it.

2 B is called *the conjugate axis of the hyperbola*, and is laid off upon the line BAB' drawn through the centre perpendicular to the transverse axis, taking

$$AB = AB' = B.$$

Corollary 8. If we put

$$e = \frac{c}{A} = \frac{2c}{2A} = \frac{1}{\cos. \varphi_0}, = \sec. \varphi_0,$$

e is called *the eccentricity of the hyperbola*.

We have $c = e A,$

and this, substituted in equations (76) and (77), gives

$$r = \frac{A(1 - e^2)}{1 - e \cos. \varphi} = \frac{A(e^2 - 1)}{e \cos. \varphi - 1}, \tag{80}$$

$$r = \frac{A(e^2 - 1)}{e \cos. \varphi + 1}. \tag{81}$$

If we take

$$P = FC = c - A = A(e - 1),$$

Rectangular equation of hyperbola.

---

these may be put in the form

$$r = \frac{P(1+e)}{e \cos. \varphi - 1} \tag{82}$$

$$r = \frac{P(1+e)}{e \cos. \varphi + 1}. \tag{83}$$

Corollary 9. If we draw ECE' perpendicular to CC', and make

$$CE = CE' = AF = AF' = A$$

we have

$$\cos. EAC = \cos. E'AC = \frac{AC}{AE} = \frac{c}{A},$$

$$= \cos. \varphi_0.$$

Hence $EAC = E'AC = \varphi_0,$

and $E_1AC = \pi - \varphi_0,$

$$E'_1AC = \pi + \varphi_0.$$

$$CE = \sqrt{(AE^2 - AC^2)} = \sqrt{(A^2 - c^2)} = B.$$

Corollary 10. The equation of the hyperbola may be referred to rectangular coördinates by arts. 59 and 60. But since equation (76) differs from the equation (64) of the ellipse only in regard to the value of c, this equation may be referred to the rectangular axes CAC and BAB', by the very same formulas as in corollary 6 upon the ellipse, and we shall have

$$(A^2 - c^2) x^2 + A^2 y^2 = A^2 (A^2 - c^2),$$

or, substituting B^2,

$$-B^2 x^2 + A^2 y^2 = -A^2 B^2, \tag{84}$$

which, divided by $-A^2 B^2$, is

$$\frac{x^2}{A^2} - \frac{y^2}{B^2} = 1. \tag{85}$$

With regard to equation (77), since it may be deduced from equation (76), by changing c into $-c$, or $-c$ into c, it may be referred to rectangular coördinates by the same process, and the corresponding result may be deduced by changing in that for (76) c into $-c$. Since, however,

$$c^2 = (-c)^2,$$

the result is the same in both cases.

Equations (84) *and* (85) *are, then, the equations of both branches of an hyperbola referred to rectangular coördinates, the centre of the hyperbola being the origin, the transverse axis being the axis of x, and the conjugate axis being the axis of y.*

Corollary 11. If we wished to find the point where the curve meets the axis of y, we should have for these points

$$x = 0,$$

so that the corresponding value of y would be

$$y = \sqrt{\left(-\frac{A^2 B^2}{A^2}\right)} = \sqrt{-B^2} = \pm B\sqrt{-1},$$

which is imaginary, and there are no such points.

Corollary 12. The equation of the hyperbola may, by art. 71, be referred to oblique axes. If the origin remains at A, the result is the same as that of corollary 9 for the ellipse, by changing B^2 into $-B^2$. By this change (70) becomes

$$(A^2 \sin^2 \alpha - B^2 \cos^2 \alpha) x_1^2 + 2(A^2 \sin.\alpha \sin.\beta - B^2 \cos.\alpha \cos.\beta) x_1 y_1$$

$$+ (A^2 \sin.^2 \beta - B^2 \cos.^2 \beta) y_1^2 = -A^2 B^2. \qquad (86)$$

Corollary 13. If α and β are so taken, that

$$A^2 \sin. \alpha \sin. \beta - B^2 \cos. \alpha \cos. \beta = 0, \qquad (87)$$

or $\qquad$ tang. α tang. $\beta = \dfrac{B^2}{A^2}$ $\qquad\qquad$ (88)

the axes are said to be *conjugate* to each other ; and equation (86) becomes

$$(A^2\sin.^2\alpha - B^2\cos.^2\alpha)x_1^2 + (A^2\sin.^2\beta - B^2\cos.^2\beta)y_1^2 = -A^2 B^2 \quad (89)$$

Corollary 14. It may be proved precisely as in corollary 11 for ellipse, that a line drawn through the centre to meet the curve at both extremities is bisected at the centre, whence it is called *a diameter*. If such a diameter is assumed for the axis of x, and if we denote it by A', we have

$$A' = \frac{AB}{\sqrt{(B^2 \cos.^2 \alpha - A^2 \sin.^2 \alpha)}}$$

and if we take

$$B' = \frac{AB}{\sqrt{(A^2 \sin.^2 \beta - B^2 \cos.^2 \beta)}},$$

we have

$$A^2 \sin.^2 \alpha - B^2 \cos.^2 \beta = -\frac{A^2 B^2}{A'^2},$$

$$A^2 \sin.^2 \beta - B^2 \cos.^2 \beta = \frac{A^2 B^2}{B'^2},$$

which substituted in (89) give, by dividing by $-A^2 B^2$,

$$\frac{x_1^2}{A'^2} - \frac{y_1^2}{B'^2} = 1, \qquad\qquad (90)$$

or

$$-B'^2 x_1^2 + A'^2 y_1^2 = -A'^2 B'^2, \qquad (91)$$

which are the equations of the hyperbola referred to conjugate diameters.

Parabola.	.	Polar equation.

5. Find the equation of the locus of all the points in a plane so situated, that the distance of each of them from a given point is equal to its distance from a given line. This locus is called *the parabola*, the given point *its focus*, and the given line *its directrix*.

Solution. Let the given point be (fig. 37.) assumed as the origin of polar coördinates, and let the perpendicular AF to the given line EQ be produced to X, and let FX be the polar axis. Let

$$BF = 2\,P.$$

Draw the perpendicular MP; we have

$$r = FM = QM = BP$$

$$FP = r\ \cos.\ \varphi,$$

so that

$$r\ \cos.\ \varphi = BP - BF = r - 2\,P$$

$$r - r\ \cos.\ \varphi = 2\,P$$

$$r = \frac{2\,P}{1 - \cos.\ \varphi}, \tag{92}$$

which is the polar equation of the parabola, the origin being the focus, and the axis the perpendicular from the directrix.

Corollary 1. If equations (67) and (92) are compared together, it is evident that (92) is what (67) becomes, when

$$e = 1.$$

Corollary 2. For the point A where the curve meets the axis we have

$$\varphi = \pi,\ \cos.\ \varphi = -1$$

$$r = FA = \tfrac{1}{2}\,2\,P = P$$

Parabola referred to rectangular axes.

The point A is called the *vertex* of the parabola, and is just as far from the focus as from the directrix.

Corollary 3. The equation of the parabola may be referred to rectangular coördinates by arts. 59 and 60. If we take the vertex A for the origin, we have

$$a = FA = P$$

$$\alpha = 0, \quad \beta = \pi,$$

$$\text{sin. } (\beta - \alpha) = 0, \quad \text{cos. } (\beta - \alpha) = -1;$$

whence (9) and (10) become

$$r = \sqrt{(x^2 + y^2 - 2Px + P^2)}$$

$$\text{tang. } \varphi = \frac{y}{x - P}$$

$$\text{cos. } \varphi = \frac{1}{\sqrt{(1 + \tan.^2 \varphi)}} = \frac{x - P}{\sqrt{(y^2 + x^2 - 2Px + P^2)}} = \frac{x - P}{r}$$

$$r \text{ cos. } \varphi = x - P.$$

Now equation (92), freed from fractions, is

$$r - r \text{ cos. } \varphi = 2P,$$

in which, if we substitute the preceding values, we have

$$\sqrt{(y^2 + x^2 - 2Px + P^2)} - x + P = 2P$$

$$\sqrt{(y^2 + x^2 - 2Px + P^2)} = P + x;$$

which squared and reduced gives

$$y^2 = 4Px; \tag{93}$$

which is the equation of the parabola referred to rectangular coördinates, the origin being the vertex, and P its distance from the focus.

Corollary 4. The equation of the parabola may be referred to oblique axes, by art. 71. If the axis of x_1 is taken parallel to x, we have

$$\alpha = 0, \ \sin. \ \alpha = 0, \ \cos. \ \alpha = 1,$$

and (27) and (28) become

$$x = a + x_1 + y_1 \cos. \beta$$
$$y = b + y_1 \sin. \beta;$$

which, substituted in (93), give

$$y_1^2 \sin.^2 \beta + (2 \, b \sin. \, \beta - 4 \, P \cos. \, \beta) \, y_1$$
$$+ \, b^2 - 4 \, P \, a = 4 \, P \, x_1. \tag{94}$$

Corollary 5. If the new origin is taken at a point A_1 upon the curve, we have, by equation (93),

$$b^2 = 4 \, P \, a,$$

which reduces (94) to

$$y_1^2 \sin.^2 \beta + (2 \, b \sin. \, \beta - 4 \, P \cos. \, \beta) \, y_1 = 4 \, P \, x_1 \tag{95}$$

Corollary 6. If the inclination β is taken so that

$$2 \, b \sin. \, \beta - 4 \, P \cos. \, \beta = 0,$$

or $$\tan. \, \beta = \frac{2 \, P}{b}, \tag{96}$$

(95) becomes

$$y_1^2 \sin.^2 \beta = 4 \, P \, x_1 \tag{97}$$

$$y_1^2 = \frac{4 \, P}{\sin.^2 \beta} \, x_1; \tag{98}$$

And if we put

$$P_1 = \frac{P}{\sin.^2 \beta}, \tag{99}$$

(98) becomes

$$y_1^2 = 4 \, P_1 \, x_1. \tag{100}$$

Prolate ellipsoid of revolution.

The axes, determined by the equation (96), are said to *be conjugate to each other, and* (100) *is the equation of the parabola referred to conjugate axes.*

6. To find the equation of the surface described by the revolution of the ellipse about its transverse axis. This surface is called that of *the prolate ellipsoid of revolution,* which is the included solid.

Solution. Let CMC' (fig. 34.) be the revolving ellipse. If we use the notation of the 3d problem and solution, and let F be the origin of the polar coördinates in the system of art. 73, and the axis of revolution the polar axis, it is evident that the distance FM of each point from the origin, or any other point of the axis, remains unchanged during the revolution of the ellipse. The value of r is then independent of θ, and depends only upon the angle φ, which it makes with the axis. Hence the equation (64) of the ellipse determines the value of r for every value of φ and every position of the revolving ellipse.

It is, then, the polar equation of the prolate ellipsoid.

Corollary 1. The equation of the ellipsoid may, by art. 77, be referred to rectangular coördinates. Thus, if in equation (64) freed from fractions we substitute

$$r = \sqrt{(x^2 + y^2 + z^2)}$$

$$r \cos. \varphi = x,$$

we have

$$A \sqrt{(x^2 + y^2 + z^2)} - c\,x = A^2 - c^2 \tag{101}$$

7

Corollary 2. This equation may, by art. 90, or 93, be referred to other rectangular axes. Thus, if the origin is changed from F to A, we have for the $a, b, c,$ of art. 93,

$$a = FA = c, \; b = 0, \; c = 0,$$

whence

$$x = x_1 + c, \; y = y_1 \; z = z_1 \, ;$$

which, substituted in (101), give

$$A\sqrt{[(x_1+c)^2+y_1^2+z_1^2]}= A^2 - c^2 + c(x_1+c) = A^2 + c\, x_1.$$

Squaring and reducing, we have

$$(A^2 - c^2) \, x_1^2 + A^2 \, (y_1^2 + z_1^2) = A^2 \, (A^2 - c^2) \, ;$$

and substituting the B^2 of corollary 4 of the ellipse

$$B^2 \, x_1^2 + A^2 \, y_1^2 + A^2 \, z_1^2 = A^2 \, B^2, \qquad (102)$$

which, divided by $A^2 \, B^2$, is

$$\frac{x_1^2}{A^2} + \frac{y_1^2}{B^2} + \frac{z_1^2}{B^2} = 1 \, ; \qquad (103)$$

which is the equation of the prolate ellipsoid of revolution referred to its centre as the origin, the axis $2\,A$ of revolution being the axis of x_1.

7. To find the equation of the surface, generated by the revolution of the ellipse, about its conjugate axis. This surface is that of the *oblate ellipsoid of revolution*.

Solution. If we take the centre A (fig. 38.) of the ellipse whose transverse axis is

$$CC' = 2\,A,$$

and conjugate axis

$$BB' = 2\,B$$

Oblate ellipsoid of revolution.

for the origin of rectangular coördinates; the equation of this ellipse is

$$\frac{x^2}{A^2} + \frac{y^2}{B^2} = 1.$$

When it revolves about the axis BB' the distances MR and AR remain unchanged. Let x_1, y_1, z_1 be the coördinates of the point M of the required surface, BAB' being the axis of x_1. We have

$$AR = x_1.$$

Now the distance of the point M from the point P is, by art. 78,

$$MR = \sqrt{[(x_1 - x_1)^2 + y_1^2 + z_1^2]} = \sqrt{(y_1^2 + z_1^2)}.$$

But MR and AR are the same with coördinates AP and MP, or x and y of the point M in the plane of the ellipse; so that, for this point,

$$y = x_1, \quad x = MR = \sqrt{(y_1^2 + z_1^2)},$$

which, substituted in the equation of the ellipse, give

$$\frac{x_1^2}{B^2} + \frac{y_1^2}{A^2} + \frac{z_1^2}{A^2} = 1 ; \qquad (104)$$

which is the equation of the oblate ellipsoid of revolution referred to its centre as the origin, the axis $2B$ of revolution being the axis of x_1.

8. To find the equation of the surface formed by the revolution of the hyperbola about either its transverse or its conjugate axis. This surface is that of the *hyperboloid of revolution.*

Solution. By reasoning exactly as in the preceding solution, we find

$$\frac{x^2}{A^2} - \frac{y^2}{B^2} - \frac{z^2}{B^2} = 1 \qquad (105)$$

for the equation of the hyperboloid of revolution referred to its centre as the origin, the transverse axis 2 A being the axis of x and also the axis of revolution,

    and we find

$$- \frac{x^2}{B^2} + \frac{y^2}{A^2} + \frac{z^2}{A^2} = 1 \qquad (106)$$

for the equation of the hyperboloid of revolution referred to its centre as origin, the conjugate axis 2 B being the axis of x, and also the axis of revolution.

9. To find the equation of the surface generated by the revolution of the parabola about its axis. This surface is that of *the paraboloid of revolution.*

Solution. By reasoning exactly as in the preceding solutions, we find

$$y^2 + z^2 = 4 P x \qquad (107)$$

for the equation of the paraboloid of revolution referred to its vertex as origin, the axis of revolution being the axis of x.

10. *To find the equation of the straight line in a plane.*

Solution. Let *AB* (fig. 39.) be the line, let any point *A* in

---

Straight line.

---

it be assumed as the origin of polar coördinates, and let the polar axis be AX, which is inclined to BA by the angle

$$BAX = \lambda.$$

For every point M of this line we have, then,

$$\varphi = MAX = \lambda \,;$$

so that $$\varphi = \lambda \qquad (108)$$

is the polar equation of a straight line, which passes through the origin, and is inclined to the axis by the angle λ.

Corollary 1. *The equation of the axis is*

$$\varphi = 0.$$

Corollary 2. The straight line may be referred to rectangular axes by art. 60, and if the axis of x is that of AX, (11) gives

$$\text{tang. } \lambda = \frac{y}{x}, \qquad (109)$$

or $$y = x \text{ tang. } \lambda \,; \qquad (110)$$

which is the equation of the straight line, which passes through the origin, and is inclined to the axis of x by the angle λ.

Corollary 3. For the axis of x

$$\lambda = 0 \,;$$

so that $$y = 0 \qquad (111)$$

is the equation of the axis of x.

7*

In like manner

$$x = 0 \qquad (112)$$

is the equation of the axis of y.

Corollary 4. The straight line may be referred to any oblique coördinates by art. 71. But since the new axes may be situated in any way whatever with regard to the former ones, the generality of the result is not diminished by limiting the original position of the line to that of the axis of y, corresponding to equation (112).

Thus, if the new origin is at the point A_1, we have

$$a = A'A = A''A_1,$$

and (27) becomes

$$x_1 \cos. \alpha + y_1 \cos. \beta = - a.$$

Now $- a$ is the value of $A''A_1$ counted from A_1, or it is the perpendicular let fall upon the line from the new origin, and if we put

$$p = - a,$$

we have

$$x_1 \cos. \alpha + y_1 \cos. \beta = p; \qquad (113)$$

which is the equation of a straight line passing at the distance p from the origin, α and β being the angles which the perpendicular to the line makes with the axes of x_1 and y_1.

Corollary 5. Equation (113) may be applied to the case in which the new axes are rectangular, when

$$\beta - \alpha = \tfrac{1}{2}\pi$$

$$\beta = \tfrac{1}{2}\pi + \alpha$$

$$\cos. \beta = - \sin. \alpha \text{ and } \cos. \alpha = - \sin. \beta,$$

Straight line.

and (113) becomes

$$\cos. \alpha . x_1 - \sin. \alpha . y_1 = p, \qquad (114)$$

or $\qquad - \sin. \beta . x_1 + \cos. \beta . y_1 = p \qquad (115)$

$$\cos. \beta . y_1 = x_1 \sin. \beta + p$$

$$y_1 = x_1 \tan. \beta + p . \sec. \beta,$$

in which β is the angle made by the line itself with the axis of x^1.

In fig. 40 let AB be the line, we have in the right triangle $A_1 PB$, formed by letting fall the perpendicular $A_1 P$,

$$A_1 P = p, \ PA_1 B = \beta$$

and, if $\qquad A_1 B = A_1 P \sec. PA_1 B = p \sec. \beta,$

$$h = A_1 B = p \sec. \beta$$

$$y_1 = x_1 \tan. \beta + h, \qquad (116)$$

which is the equation of a straight line inclined to the axis of x_1 by the angle β, and cutting the axis of y_1 at a height h above the origin.

Corollary 6.   The equation of the straight line may be obtained for any polar coördinates by art. 47, or more simply by art. 61, applied to the axis of y; in this case we have, as in corollary 4,

$$a = - p,$$

and (12) substituted in (112), gives

$$r \cos. (\varphi + a) = p, \qquad (117)$$

which is the polar equation of a straight line passing at the distance p from the origin, the perpendicular upon it being inclined to the axis by the angle — a.

11. *To find the equation of a straight line in space.*

Solution. If a point in the line is assumed as the origin, and such rectangular axes of x, y, z, that the straight line makes with them the angles λ, μ, ν, the polar equations of the line in the system of art. 87, are

$$\varphi = \lambda, \ \psi = \mu, \ \omega = \nu. \tag{118}$$

Corollary 1. It must not be forgotten that λ, μ, ν are not entirely independent of each other, but are subject to the restriction of art. 86,

$$\cos.^2 \lambda + \cos.^2 \mu + \cos.^2 \nu = 1. \tag{119}$$

Corollary 2. The equations of the axis of x are

$$\varphi = 0, \ \psi = \tfrac{1}{2}\pi, \ \omega = \tfrac{1}{2}\pi; \tag{120}$$

those of the axis of y are

$$\varphi = \tfrac{1}{2}\pi, \ \psi = 0, \ \omega = \tfrac{1}{2}\pi; \tag{121}$$

those of the axis of z are

$$\varphi = \tfrac{1}{2}\pi, \ \psi = \tfrac{1}{2}\pi, \ \omega = 0. \tag{122}$$

Corollary 3. Equations (39) become, by substituting in them the values of φ, ψ, ω (118),

$$x = r \cos. \mu$$

$$y = r \cos. \nu$$

$$z = r \cos. \lambda;$$

whence

$$r = \frac{x}{\cos. \mu} = \frac{y}{\cos. \nu} = \frac{z}{\cos. \lambda}, \tag{123}$$

which are the equations of a straight line passing

Plane.

through the origin, referred to rectangular coördinates.

Corollary 4. The equation of the straight line may be referred to any rectangular axes by equations of art. 93, by which (123) becomes

$$\frac{x_1 + a}{\cos. \lambda} = \frac{y_1 + b}{\cos. \mu} = \frac{z_1 + c}{\cos. \nu}, \qquad (124)$$

which are the equations of a straight line which passes through the point, whose coördinates are — *a,* — *b,* — *c, and is inclined to the axes by the angles* λ, μ, ν.

12. *To find the equation of the plane.*

Solution. If the plane is that of xy, we have for all its points, by art. 74,

$$z = 0, \qquad (125)$$

which is, then, the equation of the plane xy.

In the same way

$$y = 0 \qquad . \qquad (126)$$

is the equation of the plane xz; also

$$x = 0 \qquad (127)$$

is the equation of the plane yz.

Corollary 1. The plane may, by arts. 90, 91, be referred to any axes whatever. Thus, for the plane of yz, equation (40) gives, by (127),

$$x_1 \cos. \alpha + y_1 \cos. \beta + z_1 \cos. \gamma = -a = p \qquad (128)$$

which is therefore *the equation of the plane, which*

passes at the distance — a or p from the origin, and the perpendicular to which is inclined by the angles α, β, γ, to the axes x_1, y_1, z_1.

Corollary 2. If the plane passes through the origin, we have

$$p = 0,$$

and (128) becomes

$$x_1 \cos. \alpha + y_1 \cos. \beta + z_1 \cos. \gamma = 0. \qquad (129)$$

13. To find the equation of the curve described by a point in the circumference of a circular wheel, which rolls in a plane upon a given straight line. This curve is called the *cycloid*.

Solution. Let the given straight line AX (fig. 41.) be the axis of x, and let the point A, at which the given point M of the circumference touched AX, be the origin ; let

$R =$ the radius CM of the wheel,

$\theta =$ the angle MCB, by which the point M is removed from B.

Then, since the arc BM has rolled over the straight line AB, it must be equal to it in length, or, by art. 97,

$$R \theta = MB = AB.$$

Draw ME parallel to AX; the right triangle CME gives

$$CE = R \cos. \theta, \; ME = BP = R \sin. \theta,$$

whence

$$x = AP = AB - BP = R\theta - R \sin. \theta, \qquad (130)$$

$$y = PM = BE = CB - CE = R - R \cos. \theta, \qquad (131)$$

Cycloid.	Spiral.

and the elimination of θ from these two equations would give the required equation. This elimination is thus effected; (131) gives, by transposition,

$$R \cos. \theta = R - y,$$

whence

$$R \sin. \theta = \sqrt{(R^2 - R^2 \cos.^2 \theta)} = \sqrt{[R^2 - (R - y)^2]}$$
$$= \sqrt{(2 R y - y^2)};$$

which, substituted in (130), gives

$$x = R \theta - \sqrt{(2 R y - y^2)}$$

$$\theta = \frac{x + \sqrt{(2 R y - y^2)}}{R},$$

which, substituted in (131), gives

$$y = R - R \cos. \left(\frac{x + \sqrt{(2 R y - y^2)}}{R} \right) \qquad (132).$$

which is the equation of the cycloid, but is not so con-venient for use as the combination of the two equations (130) *and* (131).

14. A line revolves in a plane about a fixed point of that line, to find the equation of the curve described by a moving point in that line, which proceeds from the fixed point at such a rate, that its distance from the fixed point is proportionate to the nth power of the angle made by the revolving line with the fixed line from which it starts. This curve is called a *spiral*.

Solution. Let the fixed point A (fig. 42.) be the origin, and the fixed line AB the polar axis. Let M be the moving point

---

---

which, after the line has revolved completely round once, has arrived at M'. Let

$$R = AM',$$

we have, by condition,

$$r : R = \varphi^n : (2\pi)^n,$$

or $\qquad\qquad r\,(2\pi)^n = R\,\varphi^n, \qquad\qquad (133)$

for the equation of the spiral.

Corollary 1. If $\qquad n = 1$ and $R = 1$,

equation (133) becomes

$$2\pi r = \varphi, \qquad\qquad (134)$$

which is the equation of the spiral of Conon or of Archimedes.

Corollary 2: If $\qquad\qquad n = -1$,

(133) becomes

$$(2\pi)^{-1} r = R\,\varphi^{-1},$$

or $\qquad\qquad r\,\varphi = 2\pi R, \qquad\qquad (135)$

which is the equation of the hyperbolic spiral.

Corollary 3. If the logarithm of the distance of the point had been equal to the angle, the curve would have been *the logarithmic spiral*, and its equation

$$\varphi = \log. r. \qquad\qquad (136)$$

15. To find the equation of the right cylinder, whose base is a circle.

Solution. Let the plane of the base be that of $x\,y$. For any point whatever (fig. 43.) the coördinates of its pro-

Right cylinder.	Right cone.

jection P upon the plane of xy are x and y. But since the point P is in the circumference of the base, x and y must satisfy the equation of this circumference.

The equation of the right cylinder is then the same as that of its base, if the plane of this base is assumed as one of the coördinate planes.

Corollary. The preceding proposition is obviously general, and may be applied to any right cylinder whatever, be its base a circle, an ellipse, an hyperbola, or any other curve.

16. To find the equation of the right cone, whose base is a circle.

Solution. Let the vertex A (fig. 44.) be assumed as the origin, and let the axis of x be that of the cone, and let

$\lambda =$ the angle which the side of the cone makes with the axis.

Every radius vector, as AM, is a part of a side, and therefore

$$\varphi = \lambda \qquad (137)$$

is a property of every radius vector, *and is the polar equation of a right cone, the origin being the vertex, and λ being the angle made by the side with the axis.*

Corollary 1. The equation of the cone may, by art 88, be referred to rectangular axes, and we have

$$\cos. \; \varphi = \frac{x}{r} = \frac{x}{\sqrt{(x^2 + y^2 + z^2)}} = \cos. \; \lambda,$$

8

so that

$$x = \sqrt{(x^2 + y^2 + z^2)} \cos. \lambda \qquad (138)$$

is the required equation.

Squaring and transposing, we have

$$(y^2 + z^2) \cos.^2 \lambda = x^2 (1 - \cos.^2 \lambda) = x^2 \sin.^2 \lambda$$

$$y^2 + z^2 = x^2 \tan g.^2 \lambda, \qquad (139)$$

which is the equation of a right cone whose vertex is the origin, and axis the axis of x.

Corollary 2. The equation of the cone may be referred to any rectangular axes which have the same origin, and which make the angles α, β, γ with the axis of x, by arts. 90 – 92; for we have

$$r = \sqrt{(x^2 + y^2 + z^2)} = \sqrt{(x_1^2 + y_1^2 + z_1^2)}$$

$$x = x_1 \cos. \alpha + y_1 \cos. \beta + z_1 \cos. \gamma;$$

which, substituted in (138), give

$$x_1 \cos. \alpha + y_1 \cos. \beta + z_1 \cos. \gamma = \sqrt{(x_1^2 + y_1^2 + z_1^2)} \cos. \lambda. \quad (140)$$

Corollary 3. If the origin is changed to the point, whose coördinates are a, $b = 0$, $c = 0$, the equation becomes

$$(x_2 + a) \cos. \alpha + y_2 \cos. \beta + z_2 \cos. \gamma$$

$$= \sqrt{[(x_2 + a)^2 + y_2^2 + z_2^2]} \cos. \lambda. \qquad (141)$$

Order of curves.

CHAPTER V.

CLASSIFICATION AND CONSTRUCTION OF LOCI.

99. WHEN the equations of loci are referred to rectangular coördinates, they are divided into degrees, or orders, corresponding to the degree of their equation. Thus the locus, whose equation is of the first degree, is itself of the first degree or linear, and the same is the case with other curves.

100. *Theorem.* *The order of a curve is independent of the particular system of rectangular coördinates to which it may be referred;* that is, it is of the same order for all systems of rectangular coördinates.

Demonstration. The formulas (16, 17) or (40, 41, 42) for transforming from one system of rectangular coördinates to another are linear; so that the greatest number of the dimensions of x, y, z in any term must be the same with that of the dimensions of x_1, y_1, z_1. The degree of the equation is, therefore, the same, when expressed in terms of x_1, y_1, z_1, that it is when expressed in terms of x, y, z.

101. *Corollary.* Since the equations for transforming to oblique coördinates are also linear, the preceding proposition may be extended to them.

102. *Corollary.* The degree of the circle is, by (58),

the second; likewise that of the sphere (62); of the ellipse (69); of the hyperbola (85); of the parabola (93); of the cylinder and of the cone. The degree of the straight line (123) is the first, or it is linear, as is also that of the plane (128). The equations of the cycloid, and of the spirals, cannot be expressed without the aid of arcs, so that these curves are *transcendental*.

103. *Problem.* *To construct a locus, of which the equation is given.*

Solution. I. If the equation is that of a locus in a plane, and expressed by polar coördinates, we can, by giving successive values to φ, differing but little from each other, calculate, by means of the given equation, the corresponding values of r. As many points of the required curve may thus be determined as may be convenient, and the curve, which is drawn by the hand through these points, cannot differ much from · the required curve.

II. If the locus were in a plane and expressed by rectangular coördinates, points might be determined by calculating for assumed values of x the corresponding values of y.

III. If the equation were that of a locus in space, and expressed by polar coördinates; then for each assumed direction of the radius its value might be calculated, and the locus obtained by joining the series of points thus determined would obviously be a surface.

Construction of loci.

IV. If the equation were that of a locus in space, expressed by rectangular coördinates, values might be assumed for x and y, and the corresponding value of z would express the height at which the point of the locus was above its projection upon the plane xy; so that this locus would also be that of a surface.

V. If there were two equations in space, then one of the coördinates might be assumed at pleasure, and the corresponding values of the other two obtained.

104. *Corollary.* A single equation between coördinates in space denotes a surface. But if there are two equations, the coördinates of each point of the locus must satisfy each equation, and the point must be at once upon both the surfaces represented by these equations; so that the locus is the intersection of these surfaces, and is consequently a line.

105. In determining the character of loci from their equations, it is important that these equations should be first of all referred to those coördinates, for which they are the most simple in their forms.

CHAPTER VI.

EQUATION OF THE FIRST DEGREE.

106. THE general form of the equation of the first degree in a plane is

$$A x + B y + M = 0, \qquad (142)$$

and that of the first degree in space is

$$A x + B y + C z + M = 0. \qquad (143)$$

107. *Problem.* To reduce the general equation of the first degree in a plane to its most simple form.

Solution. Let the general formulas (16) and (17) for transformation from one system of rectangular coördinates to another in a plane be substituted in the general equation (142). The result is

$$(A \cos. \alpha + B \sin. \alpha) x_1 + (B \cos. \alpha - A \sin. \alpha) y_1$$
$$+ A a + B b + M = 0 \quad (144)$$

in which a and b are the coördinates of the new origin, and α the angle made by the axes x and x_1.

Now the position of the new origin may be assumed at such a point that

$$A a + B b + M = 0, \qquad (145)$$

and the angle α may be so assumed that

$$A \cos. \alpha + B \sin. \alpha = 0,$$

or $$\qquad \text{tang.} \, \alpha = - \frac{A}{B}, \qquad (146)$$

Linear locus.	Perpendicular.

whence

$$\cos. \alpha = \frac{1}{\sqrt{(1 + \text{tang.}^2 \alpha)}} = \frac{B}{\sqrt{A^2 + B^2})}$$

$$\sin. \alpha = \cos. \alpha . \text{tang.} \alpha = - \frac{A}{\sqrt{(A^2 + B^2)}},$$

and (144) is reduced to

$$\frac{A^2 + B^2}{\sqrt{(A^2 + B^2)}} y_1 = 0,$$

or $$\sqrt{(A^2 + B^2)} y_1 = 0,$$

whence $$y_1 = 0, \qquad (147)$$

which is as simple a form as the given equation can attain.

108. *Corollary.* Since

$$y_1 = 0$$

is, by (111), the equation of the axis of x_1, *the locus of the given equation is a straight line, which passes through the point of which the coördinates are a and b, and is inclined to the axis of x by the angle whose tangent is — A ÷ B.*

109. *Corollary.* If the given equation (142) is divided by $\sqrt{(A^2 + B^2)}$, and cos. α and sin. α are substituted for their values, it becomes, by transposition,

$$- \sin. \alpha . x + \cos. \alpha . y = - \frac{M}{\sqrt{(A^2 + B^2)}},$$

which, compared with (115), leads to the result that

$$- \frac{M}{\sqrt{(A^2 + B^2)}}$$

is the length of the perpendicular let fall upon the line from the origin.

Of the two values of

$$\sqrt{(A^2 + B^2)} = \pm\sqrt{(A^2 + B^2)},$$

that value should then be taken which renders

$$- M \div \sqrt{(A^2 + B^2)}$$

positive, that is, the value which is of the same sign with $- M$.

110. *Problem.* To find the angle of two lines in a plane, whose equations are given.

Solution. Let their equations be

$$A\, x + B\, y + M = 0,$$

$$A_1 x + B_1 y + M_1 = 0;$$

and let α, α_1 be the angles which they respectively make with the axis of x; we have for the angle I, which they make with each other,

$$I = \alpha_1 - \alpha. \tag{148}$$

But, by (146), we have

$$\text{tang. } \alpha = - \frac{A}{B}, \text{ tang. } \alpha_1 = - \frac{A_1}{B_1};$$

and

$$\text{tang. } I = \text{tang. } (\alpha_1 - \alpha)$$

$$= \frac{\text{tang. } \alpha_1 - \text{tang. } \alpha}{1 + \text{tang. } \alpha \, \text{tang. } \alpha_1}$$

or

$$\text{tang. } I = \frac{AB_1 - A_1 B}{AA_1 + BB_1}. \tag{149}$$

111. *Corollary.* If the two lines are parallel, we have

$$I = 0$$

$$\text{tang. } I = 0$$

$$A_1 B - B_1 A = 0, \tag{150}$$

$$\text{or} \quad \frac{A_1}{B_1} = \frac{A}{B} \tag{151}$$

for the equation expressing that the two lines are parallel.

112. *Corollary.* If the two lines are perpendicular, we have

$$I = \tfrac{1}{2}\pi,$$

$$\text{tang. } I = \infty = \tfrac{1}{0};$$

or the denominator of (149) must be zero; that is,

$$A A_1 + B B_1 = 0 \tag{152}$$

is the equation expressing that the two lines are perpendicular.

113. *Corollary.* In case the two lines are parallel, their distance apart must, by art. 109, be

$$-\frac{M}{\sqrt{(A^2 + B^2)}} + \frac{M_1}{\sqrt{(A_1^2 + B_1^2)}}.$$

114. *Problem.* To find the coördinates of the point of the intersection of two straight lines in a plane.

Solution. Let the coördinates of the point of intersection be x_0 and y_0, and let the equations of the line be the same as in the preceding article. Since the point of intersection is upon each line, its coördinates must satisfy each of their equations, or we must have

$$A \, x_0 + B \, y_0 + M = 0,$$
$$A_1 x_0 + B_1 y_0 + M_1 = 0;$$

from which the values of x_0 and y_0 are found to be

$$x_0 = \frac{BM_1 - B_1 M}{AB_1 - A_1 B}, \tag{153}$$

$$y_0 = \frac{A_1 M - AM_1}{AB_1 - A_1 B}; \tag{154}$$

115. *Corollary.* If the equations of the line had been given in the form corresponding to (115)

$$- \sin. \; \alpha \, . \, x + \cos. \; \alpha \, y = p$$
$$- \sin. \alpha_1 \, . \, x + \cos. \alpha_1 y = p_1$$

we should have found

$$x_0 = \frac{p \cos. \alpha_1 - p_1 \cos. \alpha}{\sin. \alpha_1 \cos. \alpha - \cos. \alpha_1 \sin. \alpha} = \frac{p \cos. \alpha_1 - p_1 \cos. \alpha}{\sin. (\alpha_1 - \alpha)} \tag{155}$$

$$y_0 = \frac{p \sin. \alpha_1 - p_1 \sin. \alpha}{\sin. (\alpha_1 - \alpha)}. \tag{156}$$

116. *Corollary.* The values of x_0 and y_0 (153) and (154) would be infinite, if their denominators were zero, that is, if we had

$$AB_1 - A_1 B = 0,$$

or by (150) if they were parallel, in which case they would not meet, and there would be no point of intersection.

117. *Problem.* To find the equation of a straight line, which makes a given angle with a given straight line.

Line inclined to given line.

Solution. Let the given angle be I, and the equation of the given straight line

$$- \sin.\, \alpha \cdot x + \cos.\, \alpha \cdot y = p,$$

and let that of the required straight line be

$$- \sin.\, \alpha_1 \cdot x + \cos.\, \alpha_1 \cdot y = p_1,$$

in which α_1 and p_1 are unknown. We have, by the conditions of the problem,

$$\alpha_1 - \alpha = I, \text{ or } \alpha_1 = I + \alpha;$$

and this value of α_1, being substituted in the equation of the required line, gives

$$- \sin.\, (I + \alpha) \cdot x + \cos.\, (I + \alpha) y = p_1 \qquad (157)$$

for the required equation, in which p_1 is indeterminate, because there is an infinite number of lines which satisfy the condition of the problem.

118. *Corollary.* If the required line is to be parallel to the given line, we have

$$I = 0,$$

and (157) becomes

$$- \sin.\, \alpha \cdot x + \cos.\, \alpha \cdot y = p_1. \qquad (158)$$

119. *Corollary.* If the required line is to be perpendicular to the given line, we have

$$I = \tfrac{1}{2}\pi, \ \sin.\, (I + \alpha) = \cos.\, \alpha, \ \cos.\, (I + \alpha) = - \sin.\, \alpha,$$

and (157) becomes

$$\cos.\, \alpha\, x + \sin,\, \alpha\, y = - p_1. \qquad (159)$$

120. *Problem.* To find the equation of a straight line, which passes through a given point.

Solution. Let x', y' by the coördinates of the given point, and

$$-\sin. \alpha . x + \cos. \alpha . y = p,$$

the required equation in which α and p are unknown. Since the given point is in the required line, its coördinates must satisfy this equation, and we have

$$-\sin. \alpha . x' + \cos. \alpha . y' = p, \qquad (160)$$

which is a condition that must be satisfied by α and p; although it is not sufficient to determine their values, because many different lines can be drawn through the same point. If the value of p is substituted in the required equation, it becomes, by transposition,

$$-\sin. \alpha . (x - x') + \cos. \alpha . (y - y') = 0, \qquad (161)$$

or, dividing by cos. α,

$$-\tan. \alpha (x - x') + (y - y') = 0, \qquad (162)$$

which is the required equation, α being indeterminate.

121. *Corollary.* If this straight line is also to pass through another point, the coördinates of which are x'' and y'', we also have this condition corresponding to (160)

$$-\sin. \alpha . x'' + \cos. \alpha . y'' = p,$$

from which and (160) the values of p and α are to be found.

The difference between the last equation and (160), divided by cos. α and by $x'' - x'$, is

$$\tan. \alpha = \frac{y'' - y'}{x'' - x'}; \qquad (163)$$

which substituted in (162) gives, by transposition,

$$y - y' = \frac{y'' - y'}{x'' - x'} (x - x') \qquad (164)$$

for the equation of a straight line, which passes through the two points whose coördinates are x', y' and x'', y''.

122. *Corollary.* If the straight line of art. 120 has also to make a given angle with the straight line whose equation is

$$- \sin. \alpha_1 \, x + \cos. \, \alpha_1 \, y = p_1,$$

we have, by art. 117,

$$a = I + \alpha_1,$$

which substituted in (162) gives, by transposition,

$$y - y' = \tan. \, (I + \alpha_1) \, (x - x') \qquad (165)$$

for the required equation.

123. *Corollary.* If the two lines of the preceding article are to be parallel, we have

$$I = 0,$$

and (165) becomes

$$y - y' = \tan. \, \alpha_1 \, (x - x'). \qquad (166)$$

124. *Corollary.* If they are to be perpendicular, we have

$$I = \tfrac{1}{2} \pi, \; \tan. \, (\tfrac{1}{2} \pi + \alpha_1) = - \cot an. \, \alpha_1,$$

and (165) becomes

$$y - y' = - \cot an. \, \alpha_1 \, (x - x'), \qquad (167)$$

which is, therefore, *the equation of the perpendicular let fall from the point, whose coördinates are x' y' upon the straight line, whose equation is*

$$- \sin. \, \alpha_1 \, x + \cos. \, \alpha_1 \, y = p_1.$$

125. Corollary. For the point of intersection of the perpendicular (167) with the line upon which it falls, that is, for the foot of the perpendicular, we find by the process of art. 114,

$$x_0 = y' \sin. \alpha_1 \cos. \alpha_1 + x' \cos.^2 \alpha_1 - p_1 \sin. \alpha_1 \quad (168)$$

$$y_0 = y' \sin.^2 \alpha_1 + x' \sin. \alpha_1 \cos. \alpha_1 + p_1 \cos. \alpha_1. \quad (169)$$

126. Corollary. The length p_0 of this perpendicular is the distance between the points x', y' and x_0, y_0, so that by equation (23)

$$p_0^2 = (x' - x_0)^2 + (y' - y_0)^2;$$

but by (168) and (169)

$$x' - x_0 = x'(1 - \cos.^2 \alpha_1) - y' \sin. \alpha_1 \cos. \alpha_1 + p_1 \sin. \alpha_1$$

$$= x' \sin.^2 \alpha_1 - y' \sin. \alpha_1 \cos. \alpha_1 + p_1 \sin. \alpha_1$$

$$= (x' \sin. \alpha_1 - y' \cos. \alpha_1 + p_1) \sin. \alpha_1$$

$$y' - y_0 = - (x' \sin. \alpha_1 - y' \cos. \alpha_1 + p_1) \cos. \alpha_1,$$

so that

$$p_0^2 = (x' \sin. \alpha_1 - y' \cos. \alpha_1 + p_1)^2 (\sin.^2 \alpha_1 + \cos.^2 \alpha_1)$$

$$= (x' \sin. \alpha_1 - y' \cos. \alpha_1 + p_1)^2$$

$$p_0 = x' \sin. \alpha_1 - y' \cos. \alpha_1 + p_1. \quad (170)$$

127. Problem. To reduce the general equation of the first degree in space to its most simple form.

Solution. Let the general formulas (40, 41, 42) for transformation from one equation of rectangular coördinates to another in space be substituted in the general equation (143)

$$A x + B y + C z + M = 0,$$

Linear locus in space.

---

the result is

$$(A \cos. \alpha + B \cos. \alpha' + C \cos. \alpha'') x_1$$

$$+ (A \cos. \beta + B \cos. \beta' + C \cos. \beta'') y_1$$

$$+ (A \cos. \gamma + B \cos. \gamma' + C \cos. \gamma'') z_1$$

$$+ A a + B b + C c + M = 0, \tag{171}$$

in which $\alpha, \beta, \gamma, \alpha', \beta', \gamma', \alpha'', \beta'', \gamma''$ are subject to the six conditions (44 – 49).

Let now the position of the new origin be assumed at such a point, that its coördinates a, b, c satisfy the equation

$$A a + B b + C c + M = 0, \tag{172}$$

and let the angles β and γ be subject to the two conditions

$$A \cos. \beta + B \cos. \beta' + C. \cos. \beta'' = 0 \tag{173}$$

$$A \cos. \gamma + B \cos. \gamma' + C. \cos. \gamma'' = 0. \tag{174}$$

By this means equation (171), divided by

$$A \cos. \alpha + B \cos. \alpha' + C \cos. \alpha'',$$

is reduced to

$$x_1 = 0. \tag{175}$$

128. *Corollary.* Let

$$A \cos. \alpha + B \cos. \alpha' + C \cos. \alpha'' = L, \tag{176}$$

and if (176) is multiplied by cos. α, (173) by cos. β, (174) by cos. γ, the sum of the products, reduced by means of equations (47, 50, 51), is

$$A = L \cos. \alpha. \tag{177}$$

In the same way we find

$$B = L \cos. \alpha' \tag{178}$$

$$C = L \cos. \alpha''. \tag{179}$$

The sum of the squares of (177, 178, 179) is, by art. 86,

$$A^2 + B^2 + C^2 = L^2 \tag{180}$$

$$L = \sqrt{(A^2 + B^2 + C^2)}, \tag{181}$$

whence

$$\cos. \alpha = \frac{A}{L} = \frac{A}{\sqrt{(A^2 + B^2 + C^2)}} \tag{182}$$

$$\cos. \alpha' = \frac{B}{L} = \frac{B}{\sqrt{(A^2 + B^2 + C^2)}} \tag{183}$$

$$\cos. \alpha'' = \frac{C}{L} = \frac{C}{\sqrt{(A^2 + B^2 + C^2)}}. \tag{184}$$

129. *Corollary.* Since

$$x_1 = 0$$

is the equation of the plane $y_1 z_1$ the locus of the general equation (143) of the first degree in space is a plane, the perpendicular to which is inclined to the axes by angles, which are determined by equations (182–184).

130. *Corollary.* Since the intersection of two planes is a straight line, *the locus of two equations of the first degree is a straight line.*

131. *Corollary.* If the equation (143) is divided by L, and the values (182–184) substituted in the result, it becomes

$$\cos. {}^{\alpha} x + \cos. \alpha' y + \cos. \alpha'' z = -\frac{M}{L},$$

Angle of planes.

which, compared with (128), leads to the conclusion that

$$-\frac{M}{L} = p$$

is the length of the perpendicular let fall upon the plane from the origin.

132. Problem. To find the angle of two planes.

Solution. Let their equations be

$$A\,x + B\,y + C\,z + M = 0$$
$$A_1 x + B_1 y + C_1 z + M_1 = 0,$$

and let $\alpha\,\beta\,\gamma$, $\alpha_1\,\beta_1\,\gamma_1$ be the angles which the perpendiculars to them make with the axes of x, y, z; and let I be the angle of the planes. The angle I is also that made by the perpendiculars to the planes, so that, by (43),

$$\cos. I = \cos. \alpha \cos. \alpha_1 + \cos. \beta \cos. \beta_1 + \cos. \gamma \cos. \gamma_1, \quad (185)$$

and by equations (181 – 184)

$$\cos. I = \frac{A A_1 + B B_1 + C C_1}{L L_1}$$

$$= \frac{A A_1 + B B_1 + C C_1}{\sqrt{(A^2 + B^2 + C^2)}.\sqrt{(A_1^2 + B_1^2 + C_1^2)}}. \quad (186)$$

133. Corollary. If the planes are parallel, their perpendiculars are parallel, and make equal angles with the axes, so that

$$\alpha = \alpha_1, \ \beta = \beta_1, \ \gamma = \gamma_1,$$

or $\qquad \dfrac{A}{L} = \dfrac{A_1}{L_1}, \ \dfrac{B}{L} = \dfrac{B_1}{L_1}, \ \dfrac{C}{L} = \dfrac{C_1}{L_1}, \qquad (187)$

9*

or $\qquad \dfrac{A}{A_1} = \dfrac{B}{B_1} = \dfrac{C}{C_1} = \dfrac{L}{L_1},$ $\qquad$ (188)

and their distance apart in this case is

$$\frac{M_1}{L_1} - \frac{M}{L}. \qquad (189)$$

134. *Corollary.* If the planes are perpendicular, we have

$$I = 90°, \quad \cos. I = 0,$$

and (185) and (186) give

$$\cos. \alpha \cos. \alpha_1 + \cos. \beta \cos. \beta_1 + \cos. \gamma \cos. \gamma_1 = 0 \quad (190)$$

$$AA_1 + BB_1 + CC_1 = 0. \qquad (191)$$

135. *Corollary.* Since

$$\sin. I = \sqrt{(1 - \cos.^2 I)}$$

we have, by (186),

$$\sin.^2 I = 1 - \frac{(AA_1 + BB_1 + CC_1)^2}{(A^2 + B^2 + C^2)(A_1^2 + B_1^2 + C_1^2)}$$

$$= \frac{(A^2 + B^2 + C^2)(A_1^2 + B_1^2 + C_1^2) - (AA_1 + BB_1 + CC_1)^2}{(A^2 + B^2 + C^2)(A_1^2 + B_1^2 + C_1^2)}$$

$$= \frac{\left\{ \begin{array}{c} A^2 B_1^2 - 2 AA_1 BB_1 + A_1^2 B^2 + A^2 C_1^2 - 2 AA_1 CC_1 \\ + A_1^2 C^2 + B^2 C_1^2 - 2 BB_1 CC_1 + B_1^2 C^2 \end{array} \right\}}{(A^2 + B^2 + C^2)(A_1^2 + B_1^2 + C_1^2)}$$

$$= \frac{(AB_1 - A_1 B)^2 + (AC_1 - A_1 C)^2 + (BC_1 - B_1 C)^2}{(A^2 + B^2 + C^2)(A_1^2 + B_1^2 + C_1^2)}. \quad (192)$$

Angle of line and plane.

We also have, by (181 – 184),

$$\sin.^2 I = \frac{(AB_1 - A_1 B)^2 + (AC_1 - A_1 C)^2 + (BC_1 - B_1 C)^2}{L^2 L_1^2}$$

$$= (\cos. \alpha \cos. \beta_1 - \cos. \beta \cos. \alpha_1)^2$$

$$+ (\cos. \alpha \cos. \gamma_1 - \cos. \gamma \cos. \alpha_1)^2$$

$$+ (\cos. \beta \cos, \gamma_1 - \cos. \gamma \cos. \beta_1)^2. \qquad (193)$$

136. *Problem.* To find the angle which a line makes with a plane.

Solution. If α, β, γ, are the angles which the line makes with the axes, and α_1, β_1, γ_1, those which the perpendicular to the plane makes, and I the angle made by the given line with the plane, the angle which the line makes with the perpendicular to the plane will be the complement of I, and we shall have

$$\sin. I = \cos. \alpha \cos. \alpha_1 + \cos. \beta \cos. \beta_1 + \cos. \gamma \cos. \gamma_1 \qquad (194)$$

$$\cos.^2 I = (\cos. \alpha \cos. \beta_1 - \cos. \beta \cos. \alpha_1)^2$$

$$+ (\cos. \alpha \cos. \gamma_1 - \cos. \gamma \cos. \alpha_1)^2$$

$$+ (\cos. \beta \cos. \gamma_1 - \cos. \gamma \cos. \beta_1)^2. \qquad (195)$$

137. *Corollary.* If the line is parallel to the plane, we have

$$\sin. I = 0 = \cos. \alpha \cos. \alpha_1 + \cos. \beta \cos. \beta_1 + \cos. \gamma \cos. \gamma_1. \qquad (196)$$

138. *Corollary.* If the line is perpendicular to the plane, we have

$$\alpha = \alpha_1, \quad \beta = \beta_1, \quad \gamma = \gamma_1.$$

139. Problem. To find the equation of a plane, which passes through a given point.

Solution. Let α, β, γ be the angles which the perpendicular to the plane makes with the axes, and let x', y' z' be the coordinates of the point, and p the perpendicular let fall upon the plane from the origin.

The equation of the plane is

$$\cos. \alpha . x + \cos. \beta . y + \cos. \gamma . z = p,$$

and since the point is in this plane, its coördinates satisfy the equations of the plane, and we have

$$\cos. \alpha . x' + \cos. \beta . y' + \cos. \gamma . z' = p;$$

and if the value of p thus obtained is substituted in the equation of the plane, it gives

$$\cos. _\alpha (x - x') + \cos. \beta (y - y') + \cos. _\gamma (z - z') = 0, \quad (197)$$

in which α, β, γ are arbitrary.

140. Corollary. The distance of this plane from another plane parallel to it, and which passes at the distance p_1 from the origin, is

$$p - p_1 = \cos. \alpha . x' + \cos. \beta . y' + \cos. \gamma . z' - p_1, \quad (198)$$

which is therefore the length of the perpendicular let fall from the point x', y', z', upon the plane, whose equation is

$$\cos _\alpha . x + \cos. \beta . y + \cos. \gamma . z = p_1.$$

141. Examples involving Linear Loci.

1. To find the locus of all the points so situated in a plane, that m times the distance of either of them from a given line,

Examples of linear loci.

added to n times its distance from another given line, is equal to a given length.

Solution. Let the first given line be the axis of x, and let the intersection of the two lines be the origin, and α the angle which these lines make with each other. Then, if x, y are the coördinates of one of the points of the locus, we have

$$y = \text{the distance from the first line,}$$

and if p_0 is the distance from the second line, we have, by (170),

$$p_0 = x \sin. \alpha - y \cos. \alpha.$$

If, then, l is the given length,

$$l = m\,y + n\,p_0$$
$$l = m\,y + n\,x \sin. \alpha - n\,y \cos. \alpha$$
$$- n . \sin. \alpha . x + (n \cos. \alpha - m)\,y = - l$$

so that the required locus is, by art. 108, a straight line, inclined to the first given line by the angle β, such that

$$\tan g. \beta = \frac{n \sin. \alpha}{n \cos. \alpha - m},$$

and which, by art. 109, passes at a distance from the intersection of the two lines equal to

$$\frac{l}{\sqrt{[n^2 \sin.^2 \alpha + (n \cos. \alpha - m)^2]}} = \frac{l}{\sqrt{(n^2 + m^2 - 2\,m\,n \cos. \alpha)}}.$$

Scholium. The whole length of the line thus obtained satisfies the algebraical conditions of the problem, but not the intended conditions. For at those points, where the value of y or that of p_0 is negative, l is no longer the absolute sum of $m\,y$, and $n\,p_0$ but their difference.

Examples of linear loci.

Corollary. When $m = n$

we have

$$\text{cotan. } \beta = \frac{\cos. \alpha - 1}{\sin. \alpha} = \frac{-2\sin.^2 \tfrac{1}{2} \alpha}{2\sin. \tfrac{1}{2} \alpha \cos. \tfrac{1}{2} \alpha} = -\text{tang. } \tfrac{1}{2} \alpha$$

$$\beta = 90° + \tfrac{1}{2} \alpha,$$

and the distance from the point of intersection becomes

$$\frac{l}{n\sqrt{(2 - 2\cos. \alpha)}} = \frac{l}{2n \sin. \tfrac{1}{2} \alpha}.$$

2. To find the locus of all the points so situated in a plane, that the difference of the squares of the distances of either of them from two given points in that plane is equal to a given surface.

Ans. If $2a =$ the distance of the two given points apart, and if the given surface is a parallelogram, whose base is a and altitude b, the required locus is a straight line, drawn perpendicular to the line joining the given points, and at a distance equal to $\tfrac{1}{4} b$ from the middle of this line.

3. To find the locus of all the points, from either of which if perpendiculars are let fall upon given planes, and if the first of these perpendiculars is multiplied by m, the second by m_1, the third by m_2, &c., the sum of the products is a given length l.

Ans. If

$$\cos. \alpha \ x + \cos. \beta \ y + \cos. \gamma \ z = p$$
$$\cos. \alpha_1 \ x + \cos. \beta_1 \ y + \cos. \gamma_1 \ z = p_1$$

&c., are the given planes, the required locus is a plane, whose equation is

$$(m\cos.\alpha + m_1\cos.\alpha_1 + \&c.) x + (m\cos.\beta + m_1\cos.\beta_1 + \&c.) y$$

$$+ (m\cos.\gamma + m_1\cos.\gamma_1 + \&c.) z = l + mp + m_1 p_1 + \&c.$$

or if the letter S. is used to denote the sum of all quantities of the same kind, so that

$$S \cdot m = m + m_1 + \&c.$$

the equation of this plane may be written

$$S \cdot m \cos. ^a . x + S \cdot m \cos. ^\beta . y + S \cdot m \cos. ^\gamma . z = l + S \cdot m \, p.$$

Scholium. This result is subject to limitations, precisely similar to those of example 1.

4. To find the locus of all the points, whose distances from several given points is such that if the square of the distance of either of them from the first given point is multiplied by m_1, that of its distance from the second given point by m_2, &c. the sum of the products is a given surface V. The quantities m_1, m_2, &c. are some of them to be negative, and subject to the limitation that their sum is zero.

Ans. If x_1, y_1, z_1 is the first given point, x_2, y_2, z_2 the second one, &c., and if S is used as in the preceding example, we have

$$S \cdot m_1 = 0,$$

and the required locus is the plane whose equation is

$$x S \cdot m_1 x_1 . + y S \cdot m_1 y_1 . + z S \cdot m_1 z_1 = S \cdot m_1 (x_1^2 + y_1^2 + z_1^2) . - V$$

CHAPTER VII.

EQUATION OF THE SECOND DEGREE.

142. THE general form of the equation of the second degree in a plane is

$$A x^2 + B x y + C y^2 + D x + E y + M = 0, \quad (199)$$

and that of the equation in space is

$$A x^2 + B x y + C y^2 + D x z + E y z + F z^2$$
$$+ H x + I y + K z + M = 0. \quad (200)$$

143. *Problem.* To reduce the general equation of the second degree in a plane to its simplest form.

Solution. I. By substituting in (199) equations (18) and (19) for transformation from one system of rectangular co-ordinates to another, the origin being the same; representing the coefficients of x_1^2, y_1^2, x_1, and y_1 by A_1, B_1, D_1, and E_1; and taking α of such a value that the coefficient of $x_1 y_1$ may be zero; (199) becomes

$$A_1 x_1^2 + B_1 y_1^2 + D_1 x_1 + E_1 y_1 + M = 0. \quad (201)$$

in which we have

$$A_1 = A \cos^2 \alpha + B \sin \alpha \cos \alpha + C \sin^2 \alpha \quad (202)$$

$$B_1 = A \sin^2 \alpha - B \sin \alpha \cos \alpha + C \cos^2 \alpha \quad (203)$$

$$D_1 = D \cos \alpha + E \sin \alpha \quad (204)$$

$$E_1 = -D \sin \alpha + E \cos \alpha, \quad (205)$$

Reduction of quadratic equation.

and α satisfies the equation

$$2(C-A)\sin.\alpha\cos.\alpha + B(\cos.^2\alpha - \sin.^2\alpha) = 0. \quad (206)$$

II. If, now, we substitute the formulas (20) for transposing the origin in (201); using x_2 and y_2 for the new coördinates; take the coördinates a and b of the new origin of such values, that the coefficients of x_2 and y_2 may be zero; and denote the sum of the terms which do not contain x_2 or y_2 by M_1; (201) becomes

$$A_1 x_2^2 + B_1 y_2^2 + M_1 = 0, \quad (207)$$

in which

$$M_1 = A_1 a^2 + B_1 b^2 + D_1 a + E_1 b + M, \quad (208)$$

and a and b satisfy the equations

$$2 A_1 a + D_1 = 0 \quad (209)$$

$$2 B_1 b + E_1 = 0. \quad (210)$$

The form (207) to which the given equation is thus reduced is its simplest form.

144. *Corollary.* If we take L, L' such that

$$L = 2 A \cos.\alpha + B \sin.\alpha \quad (211)$$

$$L' = 2 C \sin.\alpha + B \cos.\alpha, \quad (212)$$

these values may be substituted in (206), and the double of (202) would give

$$2 A_1 = L \cos.\alpha + L' \sin.\alpha \quad (213)$$

$$0 = L' \cos.\alpha - L \sin.\alpha. \quad (214)$$

The product of (213) by cos. α, diminished by that of (214) by sin. α, reduced by means of the equation

$$\sin.^2\alpha + \cos^2\alpha = 1 \quad (215)$$

10

is, by (211),

$$2 A_1 \cos. \alpha = L = 2 A \cos. \alpha + B \sin. \alpha, \qquad (216)$$

or $\qquad 2 (A_1 - A) \cos. \alpha - B \sin. \alpha = 0. \qquad (217)$

The product of (213) by sin. α added to that of (214) by cos. α is, by (215) and (212),

$$2 A_1 \sin. \alpha = L' = 2 C \sin. \alpha + B \cos. \alpha \qquad (218)$$

$$2 (A_1 - C) \sin. \alpha - B \cos. \alpha = 0. \qquad (219)$$

The product of (217) by $2 (A_1 - C)$ added to that of (219) by B is, when divided by cos. α,

$$4 (A_1 - A) (A_1 - C) - B^2 = 0, \qquad (220)$$

from which equation the value of A_1 may be determined, that is, if we put X instead of A_1, A_1 *is a root of the quadratic equation*

$$4 (X - A) (X - C) - B^2 = 0; \qquad (221)$$

the roots of which are

$$X = \tfrac{1}{2} (A + C) \pm \tfrac{1}{2} \sqrt{(B^2 + A^2 - 2 AC + C^2)}$$

$$= \tfrac{1}{2} (A + C) \pm \tfrac{1}{2} \sqrt{[B^2 + (A - C)^2]}. \qquad (222)$$

145. *Corollary.* If we take L_1 and L'_1 such that

$$L_1 = 2 A \sin. \alpha - B \cos. \alpha \qquad (223)$$

$$L'_1 = 2 C \cos. \alpha - B \sin. \alpha, \qquad (224)$$

these values may be substituted in (206) and the double of (203), and give

$$2 B_1 = L_1 \sin. \alpha + L'_1 \cos. \alpha \qquad (225)$$

$$0 = L'_1 \sin. \alpha - L_1 \cos. \alpha. \qquad (226)$$

Inclination of axis.

The product of (225) by sin. α diminished by that of (226) by cos. α is, by (215) and (223),

$$2 B_1 \sin. \alpha = L_1 = 2 A \sin. \alpha - B \cos. \alpha \qquad (227)$$

or $\qquad 2 (B_1 - A) \sin. \alpha + B \cos. \alpha = 0. \qquad (228)$

The product of (225) by cos. α added to that of (226) by sin. α is, by (215) and (224),

$$2 B_1 \cos. \alpha = L_1' = 2 C \cos. \alpha - B \sin. \alpha \qquad (229)$$

or $\qquad 2 (B_1 - C) \cos. \alpha + B \sin. \alpha = 0. \qquad (230)$

The product of (228) by $2 (B_1 - C)$ diminished by that of (230) by B, and divided by sin. α, is

$$4 (B_1 - A) (B_1 - C) - B^2 = 0, \qquad (231)$$

from which equation the value of B_1 may be determined; that is, if we put X instead of B_1, B_1 is a root of the equation (221).

A_1 and B_1 *are then the two roots* (222) *of the equation* (221).

146. *Corollary.* The value of α may be obtained from the equation (217), which gives

$$\tan g. \alpha = \frac{\sin. \alpha}{\cos. \alpha} = \frac{2 (A_1 - A)}{B}, \qquad (232)$$

or it may be obtained directly from (206).

If we substitute in (206)

$$\sin. (2\alpha) = 2 \sin. \alpha \cos. \alpha, \quad \cos. 2\alpha = \cos.^2 \alpha - \sin.^2 \alpha, \qquad (233)$$

it becomes

$$(C - A) \sin. 2\alpha + B \cos. 2\alpha = 0 ; \qquad (234)$$

Case of two lines.

whence

$$\tan. 2\,\alpha = \frac{\sin. 2\,\alpha}{\cos. 2\,\alpha} = \frac{B}{A - C}. \tag{235}$$

147. *Scholium.* The values of A_1 and B_1 (222) are always real as well as that of α (235), and those of D_1 and E_1 (204) and (205); but the equations (209) (210) are impossible if A_1 and B_1 are both zero, while D_1 and E_1 are not zero, or if either A_1 or B_1 is zero, while the corresponding value D_1 or E_1 is not zero.

148. *Scholium.* The values of A_1 and B_1 cannot both be zero, for, in this case, the quadratic terms would disappear from (201), and (201) could not, then, by art. 100, be a reduced form of a quadratic equation.

149. *Scholium.* If either A_1 or B_1 were zero, the corresponding root of (221) would be zero; that is, this equation would be satisfied by the value

$$X = 0,$$

which reduces it to

$$4\,AC - B^2 = 0; \tag{236}$$

and if we take A_1 for the root which vanishes, we have, by (232),

$$\tan. \alpha = -\frac{2\,A}{B}. \tag{237}$$

But

$$\cos. \alpha = \frac{1}{\sec. \alpha} = \frac{1}{\sqrt{(1 + \tan.^2 \alpha)}}; \tag{238}$$

whence

$$\cos. \; \alpha = \frac{B}{\sqrt{(B^2 + 4\,A^2)}} \qquad (239)$$

$$\sin. \; \alpha = \cos. \; \alpha \; \tang. \; \alpha = -\frac{2\,A}{\sqrt{(B^2 + 4\,A^2)}} \qquad (240)$$

$$D_1 = \frac{D\,B - 2\,A\,E}{\sqrt{(B^2 + 4\,A^2)}}; \qquad (241)$$

so that D_1 will vanish, only when

$$D\,B = 2\,A\,E; \qquad (242)$$

and in this case (201) becomes

$$B_1\,y_1^2 + E_1\,y_1 + M = 0; \qquad (243)$$

which gives

$$y_1 = \frac{E_1 \pm \sqrt{(E_1^2 - 4\,B_1\,M)}}{2\,B_1}; \qquad (244)$$

so that the required locus is the combination of two lines drawn parallel to the axis of x_1 at the distances from it equal to these two values of y_1, unless these values are imaginary or equal, in the former of which cases there is no locus, and in the latter the given equation is the square of the equation of the line.

150. *Scholium.* If the values of A, B, C satisfy (236), so that one of the roots of (221) is zero, and if this one is taken for A_1, we have for the other root, by (222),

$$B_1 = \tfrac{1}{2}(A + C) + \tfrac{1}{2}\sqrt{(4\,AC + A^2 - 2\,AC + C^2)}$$

$$= \tfrac{1}{2}(A + C) + \tfrac{1}{2}(A + C) = A + C, \qquad (245)$$

10*

and (201) becomes

$$(A + C)y_1^2 + D_1 x_1 + E_1 y_1 + M = 0. \quad (246)$$

The origin may now be transposed as in art. 143, the co-ordinates a and b being taken of such values that the coefficient of y_2 may be zero, and the sum of the terms which do not contain x_2 or y_2 may be zero, and (246) is thus reduced to

$$(A + C)y_2^2 + D_1 x_2 = 0. \quad (247)$$

The values of a and b satisfy the equations

$$2(A + C) b + E_1 = 0 \quad (248)$$

$$(A + C) b^2 + D_1 a + E_1 b + M = 0, \quad (249)$$

whence

$$b = - \frac{E_1}{2(A + C)}$$

$$a = \frac{-(A + C) b^2 - E_1 b - M}{D_1};$$

and if we put

$$4p = - \frac{D_1}{A + C}, \quad (250)$$

(247) becomes

$$y_2^2 - 4 p x_2 = 0,$$

or

$$y_2^2 = 4 p x_2. \quad (251)$$

151. Corollary. If the equation (221) is written in the form

$$S^2 - (A + C) S + \tfrac{1}{4} (4 AC - B^2) = 0. \quad (252)$$

The term $\tfrac{1}{4} (4 AC - B^2)$ is the product of the roots A_1 and B_1 of this equation.

A_1 and B_1 are therefore of the same sign, when

Ellipse.	Point.

$4\,A\,C$ *is greater than* B^2 ; *and they are of opposite signs if* $4\,A\,C$ *is less than* B^2.

152. *Corollary.* When B^2 is less than $4\,AC$, and, consequently, A_1 and B_1 are of the same sign, we will put

$$\frac{A_1}{\pm M_1} = \frac{1}{A_2^2}, \quad \frac{B_1}{\pm M_1} = \frac{1}{B_2^2}, \tag{253}$$

that sign being prefixed to M_1, which renders the first members of these equations positive. If then (207) is divided by $\pm M_1$, the quotient is

$$\frac{x_2^2}{A_2^2} + \frac{y_2^2}{B_2^2} \pm 1 = 0. \tag{254}$$

153. *Scholium.* If M_1 were zero, the equations (253) would be absurd, but in this case equation (207) would be

$$A_1\,x_2^2 + B_1\,y_2^2 = 0 \tag{255}$$

in which both the terms of the first member have the sign, so that the equation can only be satisfied by the conditions

$$x_2 = 0, \ y_2 = 0, \tag{256}$$

which represents the origin of the axes of x_2 and y_2.

Hence, and by art. 143, *the locus of the given equation is, in this case, the point whose coördinates are the values of a and b* (209) *and* (210).

154. *Scholium.* If M_1 were of the same sign with A_1 and B_1, the upper sign would be used in equations (253) and (254), the first member of (254) would then be the sum of three positive quantities, and could not be equal to zero.

The given equation has, then, no locus, in this case.

Hyperbola.	Two lines.

155. Scholium. When M_1 is of the sign opposite to that of A_1 and B_1, the lower sign must be used in equations (253) and (254), and (254) becomes, by transposition and omitting the numbers below the letters, which are no longer necessary,

$$\frac{x^2}{A^2} + \frac{y^2}{B^2} = 1, \tag{257}$$

which is of the same form with the equation (69) *of the ellipse.*

156. Corollary. When B^2 is greater than $4\,AC$, and, consequently, A_1 and B_1 are of opposite signs, we will put

$$\frac{A_1}{\pm M_1} = \frac{1}{A_2^2}, \quad \frac{B_1}{\mp M_1} = \frac{1}{B_2^2}, \tag{258}$$

those signs being prefixed to M_1, which render the first members of these equations positive. If, then, (207) is divided by $\pm M_1$, the quotient is

$$\frac{x_2^2}{A_2^2} - \frac{y_2^2}{B_2^2} \pm 1 = 0. \tag{259}$$

157. Scholium. If M_1 were zero, the equations (258) could not be used, but in this case equation (207) would be

$$A_1 x_2^2 + B_1 y_2^2 = 0,$$

which, multiplied by A_1, gives

$$A_1^2 x_2^2 = - A_1 B_1 y_2^2 ;$$

or, extracting the root,

$$A_1 x_2 = \pm \sqrt{(- A_1 B_1)}\, y_2 ; \tag{260}$$

the second member of which is real, because A_1 and B_1 are of opposite signs.

Hyperbola.	Ellipse.

The locus of the given equation is then the combination of the two straight lines represented by the two equations included in (260), *each of which passes through the origin of* x_2 *and* y_2.

158. *Scholium.* If M_1 is not zero, equation (259) may, by omitting the numbers below the letters and transposing the terms, be written in one of the forms

$$\frac{x^2}{A^2} - \frac{y^2}{B^2} = 1 \qquad (261)$$

or
$$\frac{y^2}{B^2} - \frac{x^2}{A^2} = 1; \qquad (262)$$

and the second of these equations becomes the same as the first by changing x, y, A, B into y, x, B, A respectively.

Equation (261) *is of the same form with equation* (85) *of the hyperbola.*

159. *Theorem.* The equation (257) is necessarily that of an ellipse.

Proof. To prove this it is only necessary to show that each point of its locus is so situated, that the sum of its distances from two fixed points is always of the same length. By comparing the equation (226) with the solution of example 2, art. 98, it is apparent that, since all the points of the ellipse satisfy the equation (69), they are in the required locus; so that if all the points of the required locus are in the ellipse, the two fixed points must be in the axis of x at a distance c from the origin such that

$$c = \pm \sqrt{(A^2 - B^2)}, \qquad (263)$$

and that the given length must be $2\,A$.

Now the distance r of the point x, y from one of these fixed points is, by art. 23,

$$r = \sqrt{[(x-c)^2 + y^2]}; \qquad (264)$$

but since $y^2 = B^2 - \dfrac{B^2 x^2}{A^2}$ and $c^2 = A^2 - B^2$, we have

$$r = \sqrt{(x^2 - 2cx + c^2 + y^2)}$$

$$= \sqrt{\left(x^2 - 2cx + A^2 - \frac{B^2 x^2}{A^2}\right)}$$

$$= \sqrt{\left(\frac{A^2 - B^2}{A^2} x^2 - 2cx + A^2\right)}$$

$$= \sqrt{\left(\frac{c^2}{A^2} x^2 - 2cx + A^2\right)}$$

$$= \pm\left(\frac{cx}{A} - A\right) = \pm\frac{cx - A^2}{A} \qquad (265)$$

Now of the two signs $+$ and $-$, that must be used which gives the distance r positive. But we have

$$c < A \text{ and } x < A$$

for

$$c = \sqrt{(A^2 - B^2)}$$

and

$$x = \sqrt{\left(A^2 - \frac{A^2 y^2}{B^2}\right)}.$$

Hence

$$cx < A^2 \text{ or } cx - A^2 < 0; \qquad (266)$$

so that the lower sign must be used in (265), which gives

$$r = A - \frac{cx}{A}; \qquad (267)$$

Hyperbola.

---

so that for the distance from one of the fixed points we have

$$r_1 = A - \frac{x\sqrt{(A^2 - B^2)}}{A}, \qquad (268)$$

and for the distance from the other

$$r_2 = A + \frac{x\sqrt{(A^2 - B^2)}}{A}, \qquad (269)$$

whence $\qquad r_1 + r_2 = 2A;$ $\qquad\qquad\qquad$ (270)

that is, all the points of the required locus belong to the ellipse.

160. *Theorem.* The equation (261) is necessarily that of an hyperbola.

Proof. The proof is the same as in the preceding theorem, except that the word *difference* is to be used for *sum*, the sign of B^2 is to be changed, and in the value of r (265) the lower sign is to be used, where c and x are both positive or both negative. For, since the values of c and x are

$$c = \pm\sqrt{(A^2 + B^2)} \text{ and } x = \pm\left(A^2 + \frac{A^2 y^2}{B^2}\right)$$

we have, when c and x are of the same sign

$$cx = \sqrt{(A^2 + B^2)} \cdot \sqrt{\left(A^2 + \frac{A^2 y^2}{B^2}\right)}$$

$$cx > A^2 \text{ or } cx - A^2 > 0;$$

whence $\qquad\qquad r_1 = \frac{cx}{A} - A.$ $\qquad\qquad$ (271)

But if c and x have opposite signs the product cx is negative, so that

$$r_2 = \frac{cx}{A} + A \qquad\qquad (272)$$

whence $\qquad r_2 - r_1 = 2\,A\,;$ $\qquad\qquad$ (273)

that is, all the points of the required locus belong to the hyperbola.

161. Theorem. The equation (251) is necessarily that of a parabola.

Proof. We have only to show that the distance of each point of the locus from that point of the axis of x_2, whose distance from the origin is p, is equal to its distance from that line which is drawn parallel to the axis of y_2, and at the distance $-p$ from it. Now since the distance of the point x, y from the axis of y is x, its distance from the line parallel to it must be

$$x + p\,;$$

and its distance r from the fixed point must be

$$r = \sqrt{[(x-p)^2 + y^2]}$$
$$= \sqrt{(x^2 - 2\,p\,x + p^2 + 4\,p\,x)} = \sqrt{(x^2 + 2\,p\,x + p^2)}$$
$$= x + p, \qquad\qquad (274)$$

which is the same as the distance from the line ; all the points of the locus of equation (251) are then upon the same parabola.

162. Theorem. In different ellipses which have the same transverse axis, the ordinates which correspond to the same abscissa are proportional to the conjugate axes.

Proof. Let the common transverse axis be $2\,A$, the different conjugate axes $2\,B$, $2\,B_1$, &c., and let the ordinates,

Ratio of ordinates in ellipses and circles.

which correspond to the same abscissa x, be y, y_1 &c., we have

$$A^2 y^2 = B^2 (A_2^2 - x^2)$$
$$A^2 y_1^2 = B_1^2 (A^2 - x^2),$$

whence, by division,

$$A^2 y^2 : A^2 y_1^2 = B^2 (A_2^2 - x^2) : B_1^2 (A^2 - x^2)$$

or $\qquad y^2 : y_1^2 = B^2 : B_1^2,$

or extracting the square root

$$y : y_1 = B : B_1 = 2B : 2B_1.$$

163. *Corollary.* Since the ellipse, whose conjugate axis is equal to its transverse axis, is a circle, the ordinate of an ellipse is to the corresponding ordinate of the circle, described upon the transverse axis as a diameter, as the conjugate axis is to the transverse axis.

164. *Corollary.* In different ellipses which have the same conjugate axis, the abscissas which correspond to the same ordinate are proportional to the transverse axes.

165. *Corollary.* The abscissa of an ellipse is to the corresponding abscissa of the circle, described upon the conjugate axis as a diameter, as the transverse axis is to the conjugate axis.

166. *Corollary.* It may be proved in the same way that in different hyperbolas, which have the same transverse axis, the ordinates which correspond to the same abscissa are proportional to the conjugate axes ; and

11

that in different hyperbolas, which have the same conjugate axis, the abscissas, which correspond to the same
ordinate, are proportional to the transverse axes.

167. *Corollary.* Understanding, by an *equilateral
hyperbola*, one in which the axes are equal, the ordinate of any hyperbola is to the corresponding ordinate
of the equilateral hyperbola, described upon its transverse axis, as the conjugate axis is to the transverse
axis, and the abscissa of the hyperbola is to the corresponding abscissa of the equilateral hyperbola, described upon its conjugate axis, as the transverse axis
is to the conjugate axis.

168. The term *abscissa* is often applied, in regard to
the ellipse and hyperbola, to denote the distance of the
foot of the ordinate from either of the extremities of
the transverse axis.

Thus the abscissas of the point M (fig. 38.) of the ellipse
are
$$CP = AC - AP = A - x$$
and
$$C'P = AC + AP = A + x.$$

The abscissas of the point M (fig. 36.) of the hyperbola are
$$CP = AP - AC = x - A$$
and
$$CP = AP + AC = x + A.$$

169. *Theorem. The squares of the ordinates in an
ellipse or hyperbola are proportional to the products of
the corresponding abscissas*, the term abscissa being
used in the sense of the preceding article.

Ratio of ordinates in ellipse and hyperbola.

Proof. I. The product of the abscissas for the point x, y of the ellipse is, by the preceding article,

$$(A + x)(A - x) = A^2 - x^2 ;$$

and this product for the point x', y' is

$$A^2 - x'^2.$$

But, by equation (68), we have

$$A^2 y^2 = A^2 B^2 - B^2 x^2$$
$$A^2 y'^2 = A^2 B^2 - B^2 x'^2;$$

whence

$$A^2 y^2 : A^2 y'^2 = A^2 B^2 - B^2 x^2 : A^2 B^2 - B^2 x'^2,$$

or, reducing to lower terms,

$$y^2 : y'^2 = A^2 - x^2 : A^2 - x'^2,$$

which is the proposition to be proved.

II. In the same way, for the hyperbola, the products of the abscissas for the points x, y, and x', y' are

$$x^2 - A^2 \text{ and } x'^2 - A^2.$$

But, by equation (84),

$$A^2 y^2 = B^2 x^2 - A^2 B^2$$
$$A^2 y'^2 = B^2 x'^2 - A^2 B^2,$$

whence $$y^2 : y'^2 = x^2 - A^2 : x'^2 - A^2.$$

170. *Theorem. The squares of the ordinates in a parabola are proportional to the corresponding abscissas.*

Proof. For the point x, y we have by (93)

$$y^2 = 4 P x,$$

and for x', y' $\qquad y'^2 = 4 P x',$

whence $\qquad y^2 : y'^2 = 4 P x : 4 P x' = x : x',$

which is the proposition to be proved.

171. *Problem.* To find the magnitude of an angle, which is inscribed in a semiellipse.

Solution. Let CMC' (fig. 45.) be the semiellipse, whose semiaxes are A and B, let I be the required angle CMC', α the angle MCX, β the angle $MC'X$, and x', y' the coordinates of the point M.

Because the line MC passes through the point x', y' and the point C, whose coördinates are

$$y = 0, \quad x = AC = A,$$

we have, by art 121,

$$\text{tang. } \alpha = \frac{y'}{x' - A}; \qquad (275)$$

and, because the line MC' passes through the point $x' y'$ and the point C', whose coördinates are

$$y = 0, \quad x = -A,$$

we have

$$\text{tang. } \beta = \frac{y'}{x' + A}; \qquad (276)$$

hence

$$\text{tang. } I = \text{tang. } (\beta - \alpha) = \frac{\text{tang. } \beta - \text{tang. } \alpha}{1 + \text{tang. } \alpha \text{ tang. } \beta}$$

$$= -\frac{2 A y'}{x'^2 - A^2 + y'^2}.$$

But, by (68),

$$x'^2 = \frac{A^2}{B^2}(B^2 - y'^2);$$

and, therefore,

$$\cdots \text{tang. } I = \frac{2\,AB^2\,y'}{(A^2 - B^2)\,y'^2} = \frac{2\,AB^2}{(A^2 - B^2)\,y'}. \qquad (277)$$

172. Corollary. The product of (275) and (276) gives by the substitution of

$$y'^2 = \frac{B^2}{A^2}(A^2 - x'^2)$$

$$\text{tang. } \alpha \,.\, \text{tang. } \beta = -\frac{B^2}{A^2}, \qquad (278)$$

which is the condition that must be satisfied by the two angles α and β, in order that two lines CM and CM', drawn from the two points M and M', may meet upon the curve.

Two such lines are called *supplementary chords;* so that (278) *is the condition which expresses that two chords are supplementary.*

173. Corollary. If equation (278) is compared with (72), it is found to be identical with it; so that the condition that two chords are supplementary is identical with the condition that two diameters are conjugate.

If then a given chord, as CM, is parallel to a given diameter $B_1AB'_1$, the chord $C'M$, supplementary to CM, is parallel to the diameter $C_1AC'_1$, conjugate to $B_1AB'_1$.

174. Problem. *To draw a diameter, which is conjugate to a given diameter.*

11*

Solution. Let $B_1AB'_1$ (fig. 45.) be the given diameter. Through C draw the chord CM parallel to $B_1AB'_1$; join $C'M$ and the diameter $C_1AC'_1$, which is drawn parallel to $C'M$, is, by the preceding article, the required diameter.

175. *Problem*. To find the magnitude of the angle formed by two chords drawn from a point of the hyperbola to the extremities of its transverse axis, which are called *supplementary chords*.

Solution. The solution is the same as that of art. 177, except in regard to the sign of B^2, which being changed gives for the required angle I

$$\text{tang. } I = -\frac{2AB^2}{(A^2 + B^2)y'}. \tag{279}$$

176. *Corollary*. The corollaries of arts. 172, 173, and the construction of art. 174, may then be applied to the hyperbola, and *equation* (88) *is the condition that two chords are supplementary.*

177. *Theorem*. The chords which are drawn parallel to the conjugate of any diameter of an ellipse or hyperbola are bisected by it.

Proof. For each value of x there are two equal values of y, one positive the other negative, which are, in the ellipse,

$$y = \pm \frac{B}{A} \sqrt{(A^2 - x^2)}$$

and, in the hyperbola,

$$y = \pm \frac{B}{A} \sqrt{(x^2 - A^2)};$$

Parameter.
---

so that if for the value of x equal to AP (fig. 46.), the line MPM' is drawn parallel to the conjugate diameter, and if PM, PM' are taken each equal to the absolute value of y, the points M, M' are upon the curve, and the chord MM', which joins these points, is bisected at P.

178. *Corollary.* The same proposition and proof may be applied to the parabola, using the word axis instead of diameter.

179. *Corollary.* The chords drawn perpendicular to either axis of an ellipse or hyperbola, or to the transverse axis of the parabola are bisected by this axis.

180. *Problem.* To find the length of the chord drawn through the focus of the ellipse, the hyperbola or the parabola, perpendicular to the transverse axis; this chord is called the *parameter of the curve.*

Solution. I. Represent the parameter of the ellipse by $4p$; and its half or the ordinate is $2p$, the corresponding abscissa being, by example 3, art. 98,

$$c = \sqrt{(A^2 - B^2)} \text{ or } c^2 = A^2 - B^2.$$

Hence the equation of the ellipse gives

$$2Ap = B\sqrt{(A^2 - c^2)} = B^2$$

$$2p = \frac{B^2}{A}, \ 4p = \frac{2B^2}{A}.$$

II. In the same way in the hyperbola we should find the same values of $2p$ and $4p$.

III. In the parabola whose equation is

$$y^2 = 4\,p\,x$$

the abscissa for the parameter is p; at which point

$$y^2 = 4\,p^2, \quad y = 2\,p$$

$$\text{parameter} = 4\,p.$$

181. *Corollary.* In the ellipse or hyperbola, we have

$$A : B = B : 2\,p$$

or $2\,A : 2\,B = 2\,B : 4\,p;$

so that *the parameter is a third proportional to the transverse and conjugate axes.*

182. *Theorem. The line drawn through either extremity of a diameter of the ellipse or hyperbola, parallel to the conjugate diameter, is a tangent to the curve.*

Proof. For the two values of y are equal to zero at the point, so that either of these lines has only one point in common with the curve.

183. *Problem. To draw a tangent to the ellipse or hyperbola at a given point of the curve.*

Solution. Join the given point M to the centre A. Through the extremity C of the transverse axis draw the chord CM' parallel to AM. Join $C'M$, and the line drawn through the, parallel to $C'M$ is, by arts. 179 and 175, the required tangent.

184. *Scholium.* The drawing of tangents to these curves will be more fully treated of in a subsequent chapter.

Reduction of quadratic equation in space.

---

185. Problem. To reduce the general equation of the second degree in space to its simplest form.

Solution. I. Substitute the equations (40, 41, 42) in (200), making

$$a = 0, \quad b = 0, \quad c = 0;$$

so that the direction of the axes may be changed without changing the origin.

If we represent the coefficients of x_1^2, y_1^2, z_1^2, x_1, y_1, z_1 by

$$A_1 = A \cos^2 \alpha + B \cos. \alpha \cos. \alpha' + C \cos^2 \alpha'$$
$$+ D \cos. \alpha \cos. \alpha'' + E \cos. \alpha' \cos. \alpha'' + F \cos^2 \alpha''$$

$$B_1 = A \cos^2 \beta + B \cos. \beta \cos. \beta' + C \cos^2 \beta'$$
$$+ D \cos. \beta \cos. \beta'' + E \cos. \beta' \cos. \beta'' + F \cos^2 \beta''$$

$$C_1 = A \cos^2 \gamma + B \cos. \gamma \cos. \gamma' + C \cos^2 \gamma'$$
$$+ D \cos. \gamma \cos. \gamma'' + E \cos. \gamma' \cos. \gamma'' + F \cos^2 \gamma''$$

$$H_1 = H \cos. \alpha + I \cos. \alpha' + K \cos. \alpha''$$

$$I_1 = H \cos. \beta + I \cos. \beta' + K \cos. \beta''$$

$$K_1 = H \cos. \gamma + I \cos. \gamma' + K \cos. \gamma''$$

and take α, β, γ, α', β', γ', α'', β'', γ'' to reduce the coefficients of $x_1 y_1, x_1 z_1, y_1 z_1$ to zero; that is, to satisfy the equations

$$0 = 2 A \cos. \alpha \cos. \beta + 2 C \cos. \alpha' \cos. \beta' + 2 F \cos. \alpha'' \cos. \beta''$$

$$+ B(\cos. \alpha \cos. \beta' + \cos. \alpha' \cos. \beta) + D(\cos. \alpha \cos. \beta'' + \cos. \alpha'' \cos. \beta)$$

$$+ E (\cos. \alpha' \cos. \beta'' + \cos. \alpha'' \cos. \beta') \qquad (280)$$

$$0 = 2 A \cos. \alpha \cos. \gamma + 2 C \cos. \alpha' \cos. \gamma' + 2 F \cos. \alpha'' \cos. \gamma''$$

$$+ B(\cos. \alpha \cos. \gamma' + \cos. \alpha' \cos. \gamma) + D(\cos. \alpha \cos. \gamma'' + \cos. \alpha'' \cos. \gamma)$$

$$+ E (\cos. \alpha' \cos. \gamma'' + \cos. \alpha'' \cos. \gamma') \qquad (281)$$

$$0 = 2\,A\cos.\,\beta\cos.\,\gamma + 2\,C\cos.\,\beta'\cos.\,\gamma' + 2\,E\cos.\,\beta''\cos.\,\gamma''$$

$$+ B(\cos.\beta\cos.\gamma' + \cos.\beta'\cos.\gamma) + D(\cos.\beta\cos.\gamma'' + \cos.\beta''\cos.\gamma)$$

$$+ E\,(\cos.\,\beta'\cos.\,\gamma'' + \cos.\,\beta''\cos.\,\gamma') \tag{282}$$

which, combined with the six equations (44 – 49), completely determine the values of these quantities, equation (200) becomes

$$A_1 x_1^2 + B_1 y_1^2 + C_1 z_1^2 + H_1 x_1 + I_1 y_1 + K_1 z_1 + M = 0. \tag{283}$$

II. Substitute the equations (53 – 55) for changing the origin to the axes x_2, y_2, z_2, and (283) becomes

$$A_1\,x_2^2 + B_1\,y_2^2 + C_1\,z_2^2 + (2\,A_1\,a + H_1)\,x_2$$

$$+ (2\,B_1\,b + I_1)\,y_2 + (2\,C_1\,c + K_1)\,z_1 + M_1 = 0, \tag{284}$$

$$M_1 = A_1\,a^2 + B_1\,b^2 + C_1\,c^2 + H_1\,a + I_1\,b + K_1\,c + M,$$

and in which if a, b, c are taken to satisfy the equations

$$2\,A_1\,a + H_1 = 0 \tag{285}$$

$$2\,B_1\,b + I_1 = 0 \tag{286}$$

$$2\,C_1\,c + K_1 = 0 \tag{287}$$

(284) becomes

$$A_1\,x_2^2 + B_1\,y_2^2 + C_1\,z_2^2 + M_1 = 0. \tag{288}$$

186. *Corollary.* If we take

$$L = 2\,A\cos.\,\alpha + B\cos.\,\alpha' + D\cos.\,\alpha'' \tag{289}$$

$$L' = 2\,C\cos.\,\alpha' + B\cos.\,\alpha + E\cos.\,\alpha'' \tag{290}$$

$$L'' = 2\,F\cos.\,\alpha'' + D\cos.\,\alpha + E\cos.\,\alpha'. \tag{291}$$

These values may be substituted in (280), (281), and the double of the value of A_1, and they give

---

Quadratic equation in space.

$$2\,A_1 = L\cos.a + L'\cos.a' + L''\cos.a'' \qquad (292)$$

$$0 = L\cos.\beta + L'\cos.\beta' + L''\cos.\beta'' \qquad (293)$$

$$0 = L\cos.\gamma + L'\cos.\gamma' + L''\cos.\gamma''. \qquad (294)$$

If (292) is multiplied by cos. α, (293) by cos. β, and (294) by cos. γ, the coefficient of L in the sum of the products is by (47) unity, while those of L' and L'' are by (50) and (51) zero, so that this sum is by (289)

$$2\,A_1\cos.\alpha = L = 2\,A\cos.a + B\cos.a' + D\cos.a'', \qquad (295)$$

or $\quad 2\,(A_1 - A)\cos.\alpha - B\cos.a' - D\cos.a'' = 0. \qquad (296)$

If (292) is multiplied by cos. a', (293) by cos. β', and (294) by cos. γ', the sum of the products is, by (48, 50, 52, 290),

$$2\,A_1\cos.a' = L' = 2\,C\cos.a' + B\cos.a + E\cos.a'' \qquad (297)$$

or $\quad 2\,(A_1 - C)\cos.a' - B\cos.a - E\cos.a'' = 0. \qquad (298)$

If (292) is multiplied by cos. a'', (293) by cos. β'', and (294) by cos. γ'', the sum of the products is, by (49, 51, 52, 291),

$$2\,A_1\cos.a'' = L'' = 2\,F\cos.a'' + D\cos.a + E\cos.a' \qquad (299)$$

or $\quad 2\,(A_1 - F)\cos.a'' - D\cos.a - E\cos a' = 0. \qquad (300)$

If (296) is multiplied by $4\,(A_1 - C)\,(A_1 - F) - E^2$, (298) by $2\,B\,(A_1 - F) + DE$, (300) by $2\,D\,(A_1 - C) + BE$, the sum of the products divided by $2\cos.a$ is

$$4\,(A_1 - A)\,(A_1 - C)\,(A_1 - F) - E^2\,(A_1 - A)$$
$$- B^2\,(A_1 - F) - D^2\,(A_1 - C) - BDE = 0 \qquad (301)$$

from which the value of A_1 may be found.

187. *Corollary.* Since the value of B_1 is obtained from that of A_1 by changing $\alpha, \alpha', \alpha''$ into β, β', β'', and since by this same change and that of β, β', β'' into $\gamma, \gamma', \gamma''$, and also by that of $\gamma, \gamma', \gamma''$ into $\alpha, \alpha', \alpha''$, (280) is changed into (282), and (281) into (280); it follows that these same changes may be made in the equations from (295) to (301), and (301) will become

$$4(B_1 - A)(B_1 - C)(B_1 - F) - E^2(B_1 - A)$$
$$- B^2(B_1 - F) - D^2(B_1 - C) - BDE = 0 \quad (302)$$

from which B_1 may be found.

188. *Corollary.* Since the value of C_1 is obtained from that of B_1, by making the same changes as in the preceding article, and since, by these changes (282) is changed into (281), and (280) into (282); it follows that these changes may also be made in the equations obtained by the preceding article, (302) will thus become

$$4(C_1 - A)(C_1 - C)(C_1 - F) - E^2(C_1 - A)$$
$$- B^2(C_1 - F) - D^2(C_1 - C) - BDE = 0 \quad (303)$$

from which C_1 may be found.

189. *Corollary.* Since the equations for determining A_1, B_1, C_1 differ only in the letters which denote the unknown quantities, and since these equations are of the third degree, it is evident *that A_1, B_1, C_1 are the three roots of the equation of the third degree*

$$4(X - A)(X - C)(X - F) - E^2(X - A)$$
$$- B^2(X - F) - D^2(X - C) - BDE = 0. \quad (304)$$

190. *Scholium.* Every equation of the third degree has at least one real root, so that one at least of the three quantities A_1, B_1, C_1 must be real. If we assume this one to be A_1, the corresponding values of cos. α, cos. α', cos. α'', asdetermined by equations (296, 298, 300), and the 1st of art. 90, are also real ; so that equations (280) and (281) are satisfied without assigning any values to β, β', β'', γ, γ', γ''. If (282) is not also satisfied, let its second member be represented by D_1, and equation (200), instead of being reduced to the form (283), will become

$$A_1 x_1^2 + B_1 y_1^2 + C_1 z_1^2 + D_1 y_1 z_1$$
$$+ H_1 x_1 + I_1 y_1 + K_1 z_1 + M = 0.$$

If now the same transformation is effected upon this equation, so as to transform it to the axes of x_2, y_2, z_2, the equation for determining A_2, B_2, C_2 would be obtained from (304), by changing A, B, C, D, E, F into A_1, 0, B_1, 0, D_1, C_1, (304) thus becomes

$$4(X - A_1)(X - B_1)(X - C_1) - D_1^2(X - A_1) = 0 \quad (305)$$

the roots of which are

$$X = A_1,$$

and $X = \frac{1}{2}(B_1 + C_1) \pm \frac{1}{2}\sqrt{[D_1^2 + (B_1 - C_1)^2]}$ (306)

which are all real, so that the given equation can always be transformed to the form (283), and all the roots of (304) will be real.

191. *Scholium.* If either A_1, B_1, or C_1 is zero, one of the equations (285 – 287) is impossible, unless the corresponding value of H_1, I_1, or K_1 is zero.

192. *Scholium.* The three roots A_1, B_1, C_1 cannot all be

12

zero at the same time; for in this case (283) would be linear, and would not be a reduced form of a quadratic equation.

193. *Scholium.* If A_1 and H_1 are both zero, the values of b and c can be taken to satisfy equations (286) and (287), and (283) is then reduced to

$$B_1 y_2^2 + C_1 z_2^2 + M_1 = 0. \qquad (307)$$

194. *Scholium.* If A_1 is zero and H_1 is not so, b and c can satisfy equations (286) and (287), and a can be taken to satisfy the equation

$$M_1 = 0,$$

so that (283) is then reduced to

$$B_1 y_2^2 + C_1 z_2^2 + H_1 x_1 = 0. \qquad (308)$$

195. *Scholium.* If A_1 and B_1 are zero, c can be taken to satisfy equation (287), and if either H_1 or I_1 is not zero, a or b can be taken to satisfy the equation

$$M_1 = 0,$$

so that (283) is then reduced to

$$C_1 z_2^2 + H_1 x_2 + I_1 y_2 = 0. \qquad (309)$$

But if both H_1 and I_1 are also zero, (283) becomes

$$C_1 z_2^2 + M_1 = 0. \qquad (310)$$

196. *Scholium.* If the values of A_1, B_1, C_1, and M_1 have all the same sign, (288) is impossible, and *there is no locus.*

197. *Corollary.* If A_1, B_1, C_1 have all the same sign, which is the reverse of M_1, let A_2, B_2, C_2 be so taken, that

$$\frac{1}{A_2^2} = -\frac{A_1}{M_1}, \quad \frac{1}{B_2^2} = -\frac{B_1}{M_1}, \quad \frac{1}{C_2^2} = -\frac{C_1}{M_1}, \qquad (311)$$

and the quotient of (288), divided by $-M_1$, is

$$\frac{x_2^2}{A_2^2} + \frac{y_2^2}{B_2^2} + \frac{z_2^2}{C_2^2} - 1 = 0. \qquad (312)$$

198. *Corollary.* If two of the quantities A_1, B_1, C_1 have the same sign with M_1, while the other one, which we will assume to be A_1, has the reverse sign, we will take

$$\frac{1}{A_2^2} = -\frac{A_1}{M_1}, \quad \frac{1}{B_2^2} = \frac{B_1}{M_1}, \quad \frac{1}{C_2^2} = \frac{C_1}{M_1}, \qquad (313)$$

and the quotient of (288), divided by $-M_1$, is

$$\frac{x_2^2}{A_2^2} - \frac{y_2^2}{B_2^2} - \frac{z_2^2}{C_2^2} - 1 = 0. \qquad (314)$$

199. *Corollary.* If of the quantities A_1, B_1, C_1, one, namely C_1, has the same sign with M_1, while the other two have the reverse sign, we will take

$$\frac{1}{A_2^2} = -\frac{A_1}{M_1}, \quad \frac{1}{B_2^2} = -\frac{B_1}{M_1}, \quad \frac{1}{C_2^2} = \frac{C_1}{M_1}, \qquad (315)$$

and the quotient of (288), divided by $-M_1$, is

$$\frac{x_2^2}{A_2^2} + \frac{y_2^2}{B_2^2} - \frac{z_2^2}{C_2^2} - 1 = 0. \qquad (316)$$

200. *Corollary.* The values of $2A_2$, $2B_2$, $2C_2$ are

called the *axes of the surface* in either of the three last articles, so that the three different values of

$$\sqrt{\left(\pm \frac{M_1}{S} \right)},$$

which are found from equation (304), are the *semi-axes*.

201. *Scholium.* If M_1 is zero, the equations (311), (313), and (315) are impossible, but in this case (288) becomes

$$A_1 x_2^2 + B_1 y_2^2 + C_1 z_2^2 = 0. \tag{317}$$

202. *Scholium.* If A_1, B_1, and C_1 have all the same sign, (317) is only satisfied by the values

$$x_2 = 0, \quad y_2 = 0, \quad z_2 = 0, \tag{318}$$

so that the origin of x_2, y_2, z_2 is in this case the required locus.

203. *Corollary.* If of the three quantities, A_1, B_1, C_1, one, as C_1, is negative, while the other two are positive, we will take

$$A_1 = \frac{1}{A_2^2}, \; B_1 = \frac{1}{B_2^2}, \; -C_1 = \frac{1}{C_2^2}, \tag{319}$$

and (317) becomes

$$\frac{x_2^2}{A_2^2} + \frac{y_2^2}{B_2^2} - \frac{z_2^2}{C_2^2} = 0. \tag{320}$$

204. The form of a surface is best investigated by examining the character of its curved sections, which

---

---

are made by different planes. The farther investigation of the surfaces, represented by quadratic equations, will, therefore, be reserved for Chapter IX.

205. EXAMPLES INVOLVING PLANE QUADRATIC LOCI.

1. To find the locus of all the points in a plane, which are so situated with regard to given points in that plane, that if the square of the distance of each point from the first given point is multiplied by m', the square of its distance from the second given point by m'', &c., the sum of the products is equal to a given surface V.

Solution. Let the given points be, respectively, x', y'; x'', y'', &c.

The distances of the point x, y from these points is given by equation (23), and we have, by the conditions of the problem and using S, as in art. 141,

$$S.m' (x - x')^2 + S.m' (y - y')^2 = V,$$

or

$$S.m'.x^2 + S.m'.y^2 - 2 S.m' x'.x - 2 S.m' y'.y$$
$$+ S.m' (x'^2 + y'^2) - V = 0.$$

This equation is already of the form 201, and may be reduced to the form (207) by making

$$a = \frac{S.m' x'}{S.m'}, \quad b = \frac{S.m' y'}{S.m'}$$

$$M_1 = \frac{[(S.m')^2 - 2] . [(S.m' x')^2 + (S' m'.y')^2]}{S.m'}$$

$$+ S.m' (x'_2 + y'_2) - V.$$

12*

We have then for the axes, by (253),

$$A_2 = \sqrt{\frac{-M_1}{S \cdot m'}} = B_2,$$

so that the locus is a circle, the coördinates of whose centre are — a and — b, and whose radius is A_2.

Corollary. — M_1 and $S \cdot m'$ must be both positive or both negative.

2. To find the locus of all the points in a plane, which are so situated with regard to given lines in the plane, that if the square of the distance of each point from the first given line is multiplied by m_1, the square of its distance from the second line by M_2, &c., the sum of the products is equal to a given surface V.

Solution. Let the given lines be respectively

$$\text{sin.} \, ^{a_1} x - \text{cos.} \, ^{a_1} y = -p_1$$

$$\text{sin.} \, ^{a_2} x - \text{cos.} \, ^{a_2} y = -p_2, \&c.$$

The distances of the point x, y of the locus from these lines is given by equation (170), and give, by the conditions of the problem, and using S as before,

$$S \cdot m_1 \, (\text{sin.} \, ^{a_1} \cdot x - \cos \, ^{a_1} \cdot y - p_1)^2 = V,$$

which, developed and compared with equations (199 – 256), give A_1 and B_1 as the roots of the equation

$$4 X^2 - 4S.m_1 X + 4S.m_1 \text{sin.}^2 ^{a_1} Sm_1 \cos_2 ^{a_1} - (Sm_1 \text{sin.}^2 ^{a_1})^2 = 0,$$

and to find a,

$$\text{tan.} 2 \, a = \frac{S \cdot m_1 \, \text{sin.} \, 2 \, a_1}{S \cdot m_1 \, \text{cos.} \, 2 \, a_1},$$

and the values of a, b, m_1 may be found by equations (208–210).

3. To find the locus of the centres of all the circles which pass through a given point, and are tangent to a given line.

Ans. A parabola of which the given point is the focus, and the given line the directrix.

4. To find the locus of the centres of all the circles, which are tangent to two given circles.

Ans. When the locus is entirely contained within one of the given circles, it is an ellipse of which the foci are the two given centres, and the transverse axis is the sum of the two given radii. Otherwise, it is an hyperbola, of which the foci are the two given centres, and the transverse axis the difference of the two given radii, if the contacts are both external or both internal, and their sum, if one of the contacts is external and the other internal; and it may be remarked, that the contact with either of the given circles is external upon one branch of the hyperbola, and internal upon the other.

CHAPTER VIII.

SIMILAR CURVES.

206. *Definition.* Two curves are said to be *similar* when they can be referred to two such systems of rectangular coördinates, that if the abscissas are taken in a given ratio, the ordinates are in the same ratio.

207. *Corollary.* If the given ratio is $m : n$, and if the co-ordinates of the first curve are x, y, the corresponding ones of the second curve must be

$$\frac{n\,x}{m}, \ \frac{n\,y}{m} \ ;$$

so that if these values are substituted for the coördinates in the equation of the second curve, the equation obtained must be that of the first curve.

208. *Theorem.* Two ellipses or two hyperbolas are similar, if the ratios of their axes are equal.

Proof. I. Let the semiaxes of the two ellipses be A, B and A', B', we have, by hypothesis,

$$A : A' = B : B',$$

and the equations of these ellipses are (68)

$$\frac{x^2}{A^2} + \frac{y^2}{B^2} = 1$$

$$\frac{x^2}{A'^2} + \frac{y^2}{B'^2} = 1 ;$$

and if, in the second equation, we take the coördinates in the ratio equal to that of the axes, that is, substitute for x, $\dfrac{A'x}{A}$,

and for y, $\dfrac{B'y}{B} = \dfrac{A'y}{A}$; it becomes identical with the first equation.

II. The same reasoning may be applied to the hyperbola; but it must be observed, that the ratios of the transverse axes must be equal to that of the conjugate axes in the two hyperbolas; and the theorem must not be applied to the case in which the ratio of the conjugate axis of the first curve to the transverse axis of the second is equal to that of the transverse axis of the first curve to the conjugate axis of the second curve.

209. *Theorem.* The radii vectores, which are drawn in the same direction to two similar curves, are in the same ratio with the corresponding coördinates.

Proof. If x and y are the coördinates for the first curve, and x' and y' the coördinates for the second curve, taken as in art. 207, we have

$$\frac{x}{y} = \frac{x'}{y'};$$

so that, by (11), the angle $\varphi - \alpha$, which determines the direction of the radius vector drawn to the point x, y, is equal to the angle which determines the direction of the radius vector drawn to the point x', y'. These radii vectores must, therefore, coincide in direction, and we have for their values

$$r = x \sec. (\varphi - \alpha), \quad r' = x' \sec. (\varphi - \alpha)$$

$$\frac{r}{r'} = \frac{x}{x'} = \frac{y}{y'}.$$

210. *Similar surfaces* may be defined in the same way as similar curves, and are subject to propositions precisely like those of arts. 207 and 209.

Similar solids are solids bounded by similar surfaces.

Section of surface by a plane.

CHAPTER IX.

PLANE SECTIONS OF SURFACES.

211. Problem. *To find the section of a surface made by a plane.*

Solution. I. If the cutting plane is one of the coördinate planes, that of xy, for instance, the points of the section are all of them in this plane, and we have, therefore, for all these points

$$z = 0,$$

so that we have *only to substitute zero for z in the equation of the surface to find the equation of the intersection with the plane of x y.* In the same way by putting

$$x = 0$$

the intersection with the plane of yz is found, and the intersection with the plane of xz is found by putting

$$y = 0.$$

II. For any other plane the intersection is found by transforming the coördinates of the surface, to a system of which the cutting plane is one of the coördinate planes. If the cutting plane is supposed to be the plane of $x_1 y_1$, we shall be obliged to put

$$z_1 = 0,$$

after substituting the equations $(40-42)$ for the transformation of coördinates. But a useless operation is avoided by putting, at once,

$$z_1 = 0$$

in the equations for transformation.

The required equation is, then, obtained by substituting in the given equation of the surface the equations

$$x = a + x_1 \cos. \alpha + y_1 \cos. \beta \qquad (321)$$

$$y = b + x_1 \cos. \alpha' + y_1 \cos. \beta' \qquad (322)$$

$$z = c + x_1 \cos. \alpha'' + y_1 \cos. \beta''. \qquad (323)$$

In which a, b, c are the coördinates of a point of the cutting plane which is the origin of x_1 and y_1, α, α', α'', and β, β', β'' are the angles which the two axes of x_1, y_1 make with the given axes.

212. Corollary. If the cutting plane is parallel to the plane of xy, the axes of x_1 and y_1 may be taken parallel to the axes of x and y, and the origin may be taken in the axis of z, so that the equations $(321-323)$ become

$$x = x_1, \ y = y_1, \ z = c. \qquad (324)$$

If the cutting plane is parallel to the plane of yz, we have in the same way

$$x = x_1, \ y = b, \ z = z_1; \qquad (325)$$

and if it is parallel to the plane of yz, we have

$$x = a, \ y = y_1, \ z = z_1. \qquad (326)$$

213. Corollary. If the cutting plane passes through the axis of x, the axis of x may be taken for that of x_1, and the

Section of quadratic surface.

origin may remain as it was. In this case equations (321 – 323) become

$$x = x_1, \; y = y_1 \cos. \beta', \; z = y_1 \sin. \beta'. \quad (327)$$

If the cutting plane passes through the axis of y, and if the axis of y is taken for that of x_1, (321 – 323) become

$$x = y_1 \cos. \beta, \; y = x_1, \; z = y_1 \sin. \beta. \quad (328)$$

If the cutting plane passes through the axis of z, and if the axis of z taken for that of x_1, (321 – 323) become

$$x = y_1 \cos. \beta, \; y = y_1 \sin. \beta, \; z = x_1. \quad (329)$$

214. *Problem. To find the section of a surface of the second degree made by a plane.*

Solution. The equation (200) is the most general equation of the surface of a second degree. It may then be regarded as the equation of the surface referred to coördinate planes, of which the plane $x\,y$ is the cutting plane. By putting

$$z = 0,$$

we have then for the required section

$$A\,x^2 + B\,x\,y + C\,y^2 + H\,x + I\,y + M = 0. \quad (330)$$

From the discussion (201 – 262), it follows that if

$$B^2 - 4\,AC < 0$$

the section, if there is one, is a point or an ellipse. But if

$$B^2 - 4\,AC = 0$$

it is a parabola, a straight line, or a combination of two parallel straight lines   But if

$$B^2 - 4\,AC > 0$$

it is an hyperbola, or a combination of two straight lines.

13

215. Corollary. For the section which is parallel to the plane of $x\,y$ at the distance c, we have by (324) putting

$$H_1 = D\,c + H \tag{331}$$

$$I_1 = E\,c + I \tag{332}$$

$$M_1 = F\,c^2 + K\,c + M \tag{333}$$

$$A\,x_1^2 + B\,x_1\,y_1 + C\,y_1^2 + H_1\,x_1 + I_1\,y_1 + M_1 = 0; \tag{334}$$

so that this section is in the same class with that made by the parallel plane of $x\,y$, so far as it depends upon the value of $B^2 - 4\,AC$.

216. The values of A_1 and B_1 depend by (220, 231) only upon those of A, B, C, so that the ratios of the semiaxes A_2 and B_2 must also depend only upon A, B, C, and be the same for all the parallel sections of the quadratic locus.

Hence, *if one of the sections of a quadratic locus is an ellipse, all the parallel sections must be similar ellipses, except those which are points.*

If one of the sections is an hyperbola, all the curved parallel sections are hyperbolas; and all those sections are similar whose greater axes are transverse; and also those are similar whose greater axes are conjugate.

If one of the sections is a parabola, all the curved sections which are parallel to it are parabolas.

In all the parallel sections the axes are parallel.

217. Problem. To investigate the form of the surface of equation (312).

Solution. The numbers below the letters were only used to distinguish the different axes of coördinates; they may, then, be omitted, and (312) may be written

$$\frac{x^2}{A^2} + \frac{y^2}{B^2} + \frac{z^2}{C^2} - 1 = 0. \tag{335}$$

Ellipsoid.

I. The equation of the section parallel to the plane of $x\,y$ at the distance c from the origin is

$$\frac{x_1^2}{A^2} + \frac{y_1^2}{B^2} + \frac{c^2}{C^2} - 1 = 0, \qquad (336)$$

which is impossible when

$$c^2 > C^2$$

it is the point

$$x_1 = 0, \; y_1 = 0$$

when $\qquad\qquad c = C,$

and it is the ellipse whose semiaxes are

$$\frac{A}{C}\sqrt{(C^2 - c^2)}, \; \frac{B}{C}\sqrt{(C^2 - c^2)} \qquad (337)$$

when $\qquad\qquad c^2 < C^2.$

II. The sections parallel to the planes of $x\,z$ and $y\,z$ are easily found in the same way, and it is evident that the surface is included by eight planes, of which two are drawn parallel to the plane of $x\,y$ at the distances $+\,C$ and $-\,C$, two parallel to the plane of $x\,z$ at the distances $+\,B$ and $-\,B$, and two parallel to the plane of $y\,z$ at the distances $+\,A$ and $-\,A$.

III. The section made by any other plane must then be limited, and must therefore be an ellipse or a point, so that this surface is called that of *an ellipsoid*, whose semiaxes are A, B, C.

IV. The section made by a plane passing through the axis of x, and inclined by an angle β' to the axis of y is, by (327),

$$\frac{x_1^2}{A^2} + \left(\frac{\cos^2 \beta'}{B^2} + \frac{\sin^2 \beta'}{C^2}\right) y_1^2 - 1 = 0. \qquad (338)$$

It is, therefore, an ellipse, whose semiaxes are A and

$$1 \div \sqrt{\left(\frac{\cos^2 \beta'}{B^2} + \frac{\sin^2 \beta'}{C^2}\right)} = \frac{BC}{\sqrt{(C^2 \cos^2 \beta' + B^2 \sin^2 \beta')}}, \quad (339)$$

or if we substitute for $\cos^2 \beta'$ its value, the second semiaxis is

$$\frac{BC}{\sqrt{[C^2 + (B^2 - C^2)\sin^2 \beta']}}. \quad (340)$$

The ellipsoid may, then, be considered as generated by the revolution of an ellipse about the axis of x, the semiaxis of the ellipse, which corresponds to the axis of x, remaining constantly A, and the other axis changing from B to the value (339).

The sections made by planes passing through the axes of y and z may be found in the same way.

218. *Corollary.* If we have

$$B = C$$

the semiaxis (339) becomes equal to B, so that the ellipse retains the same value of its second axis as well as of its first, during its revolution; and the ellipsoid is one of revolution. The sections made by planes parallel to the plane of $y z$ are, in this case, circles.

219. *Corollary.* If we have

$$A = B = C$$

the revolving ellipse is a circle, and the surface is that of a sphere.

220. *Problem.* To investigate the form of the surface of equation (314).

Hyperboloid.

Solution. By omitting the numbers below the letters, (314)
may be written

$$\frac{x^2}{A_2} - \frac{y^2}{B^2} - \frac{z^2}{C^2} - 1 = 0. \tag{341}$$

I. The section, parallel to the plane of $x\,y$, at the distance
c from the origin is, by (253, 261), an hyperbola, of which
the semitransverse axis, which is parallel to the axis of x, is

$$\frac{A}{C}\sqrt{(c^2 + C^2)}, \text{ and the semiconjugate is } \frac{B}{C}\sqrt{(c^2 + C^2)}. \tag{342}$$

II. The section, parallel to the plane of $x\,z$, at the distance
b from the origin, is an hyperbola, of which the semitransverse
axis is parallel to the axis, and is

$$\frac{A}{B}\sqrt{(b^2 + B^2)}, \text{ and the semiconjugate is } \frac{C}{B}\sqrt{(b^2 + B^2)}. \tag{343}$$

III. The equation of the section, parallel to the plane of
$y\,z$, at the distance a from the origin, is, by reversing the
signs,

$$\frac{y^2}{B^2} + \frac{z^2}{C^2} + 1 - \frac{a^2}{A^2} = 0; \tag{344}$$

so that when $a^2 < A^2$

the section is imaginary, that is, none of the surface is contained between the two planes drawn parallel to the plane of
$y\,z$, at the distances $+ A$ and $- A$; so that the surface consists of two entirely distinct branches, similar to the two
branches of an hyperbola.

When $a^2 = A^2$

the section is reduced to the point

$$y = 0, \ z = 0;$$

when $a^2 > A^2$

13*

the section is an ellipse, of which the two semiaxes are

$$\frac{B}{A}\sqrt{(a^2 - A^2)} \text{ and } \frac{C}{A}\sqrt{(a^2 - A^2)}. \qquad (345)$$

IV. The section made by any plane, which cuts both branch-es, is evidently an hyperbola, for no other curve of the second degree is composed of two branches. The section made by a plane, which cuts entirely across either branch without cut-ting the other, is an ellipse ; for this is the only curve of the second degree, which returns into itself, so as to enclose a space. The section made by a plane, which cuts one branch without entirely cutting across it, and without cutting the other branch, is a parabola ; for this is the only endless curve of the second degree, which consists of only a single branch. This surface is called that of an *hyperboloid of two branches.*

V. The equation of the section, made by a plane passing through the axis of x and inclined, by an angle β', to the axis of y is, by (327),

$$\frac{x_1^2}{A^2} - \left(\frac{\cos^2\beta'}{B^2} + \frac{\sin^2\beta'}{C^2}\right) y_1^2 - 1 = 0. \qquad (346)$$

It is, therefore, an hyperbola, whose semitransverse axis, di-rected along the axis of x_1 is A, and whose semiconjugate axis is precisely that of (340). This hyperboloid may, then, be regarded as generated by the revolution of an hyperbola about the axis of x, the semitransverse axis remaining constant, and the semiconjugate axis changing in such a way, that its extremity describes an ellipse, whose semiaxes are B and C.

VI. The section, made by a plane passing through the axis of y, and inclined by an angle β to the axis of x, is, by (328),

$$\left(\frac{\cos^2\beta}{A^2} - \frac{\sin^2\beta}{C^2}\right) y_1^2 - \frac{x_1^2}{B^2} - 1 = 0. \qquad (347)$$

Hyperboloid of two branches.

When, therefore,

$$\frac{\cos.^2\beta}{A^2} - \frac{\sin.^2\beta}{C^2} > 0$$

or $$\tan g.^2\beta < \frac{C^2}{A^2} \qquad (348)$$

the section is an hyperbola, of which the semitransverse axis is

$$\frac{AC}{\sqrt{(C^2\cos.^2\beta - A^2\sin.^2\beta)}}, \text{ and the semiconjugate is } B. \quad (349)$$

When $$\tan g.^2\beta = \frac{C^2}{A^2}$$

the section is impossible, but every parallel section is a parabola.

When $$\tan g.^2\beta > \frac{C^2}{A^2}$$

the section is impossible, but there are parallel sections which are ellipses.

In the same way and with like results, the sections may be found made by planes passing through the axis of z.

221. *Corollary.* If we have

$$B = C$$

the semiaxis (340) becomes equal to B, so that the revolving hyperbola retains the same value of its second axis as at first, and the hyperboloid is one of revolution about the transverse axis. The sections made by planes parallel to the plane of $y\,z$ are, in this case, circles.

222. *Problem.* To investigate the form of the surface of equation (316).

Solution. By omitting the numbers below the letters, (316) may be written

$$\frac{x^2}{A^2} + \frac{y^2}{B^2} - \frac{z^2}{C^2} - 1 = 0. \qquad (350)$$

I. The section, made by a plane parallel to the plane of xy, at the distance c, is an ellipse, of which the semiaxes are

$$\frac{A}{C}\sqrt{(c^2 + C^2)} \text{ and } \frac{B}{C}\sqrt{(c^2 + C^2)}. \qquad (351)$$

II. The section, made by a plane parallel to the plane of xz, at the distance b, is when

$$b^2 < B^2$$

an hyperbola, of which the transverse semiaxis is parallel to the axis of x, and is

$$\frac{A}{B}\sqrt{(B^2 - b^2)}, \text{ and the semiconjugate is } \frac{C}{B}\sqrt{(B^2 - b^2)}. (352)$$

When $\qquad\qquad b^2 = B^2$

the section is the combination of the two straight lines

$$\frac{x}{A} = \pm \frac{z}{C}. \qquad (353)$$

When $\qquad\qquad b^2 > B^2$

the section is an hyperbola, whose semitransverse axis is parallel to the axis of z, and is

$$\frac{C}{B}\sqrt{(b^2 - B^2)}, \text{ the semiconjugate is } \frac{A}{B}\sqrt{(b^2 - B^2)}. (354)$$

In the same way and with like results, the sections may be found made by a plane parallel to that of yz.

Hyperboloid.

III. The curved section made by any other plane is an hyperbola when it consists of two branches, an ellipse when it is limited, and a parabola when it consists of one infinite branch.

IV. The equation of the section, made by a plane passing through the axis of x and inclined, by an angle β', to the axis of y, is

$$\frac{x_1^2}{A^2} + \left(\frac{\cos.^2 \beta'}{B^2} - \frac{\sin.^2 \beta'}{C^2} \right) y_1^2 - 1 = 0. \qquad (355)$$

When
$$\frac{\cos.^2 \beta'}{B^2} - \frac{\sin.^2 \beta'}{C^2} > 0$$

or
$$\tan g.^2 \beta' < \frac{C^2}{B^2} \qquad (356)$$

this section is that of an ellipse, whose semiaxes are

A and
$$\frac{BC}{\sqrt{(C^2 \cos.^2 \beta' - B^2 \sin.^2 \beta')}}. \qquad (357)$$

When
$$\tan g.^2 \beta' = \frac{C^2}{B^2} \qquad (358)$$

the section is reduced to the two parallel straight lines

$$x' = \pm A$$

drawn parallel to the axis of y_1. Any parallel section to this one is a parabola.

When
$$\tan g.^2 \beta' > \frac{C^2}{B^2} \qquad (359)$$

the section is an hyperbola, whose transverse semiaxis is in the direction of the axis of x, and is

A, the semiconjugate is
$$\frac{BC}{\sqrt{(B^2 \sin.^2 \beta' - C^2 \cos.^2 \beta')}} \qquad (360)$$

In the same way, and with like results, the sections may be found, made by a plane passing through the axis of y.

V. The equation of the section, made by a plane passing through the axis of z, and inclined, by an angle β to the axis of x, is by (329)

$$\left(\frac{\cos.^2\beta}{A^2} + \frac{\sin.^2\beta}{B^2}\right) y_1^2 - \frac{x_1^2}{C^2} - 1 = 0. \qquad (361)$$

This section is, therefore, an hyperbola, whose semiconjugate axis directed according to the axis of z or x_1, is

C, while the semitransverse is $\dfrac{AB}{\sqrt{(B^2\cos.^2\beta + A^2\sin.^2\beta)}}$ (362)

or the semitransverse axis is $\dfrac{AB}{\sqrt{[B^2 + (A^2 - B^2)\sin.^2\beta]}}$. (363)

This hyperboloid may, then, be regarded as generated by the revolution of an hyperbola about its conjugate axis C, the extremity of the transverse axis describing the ellipse, whose semiaxes are A and B.

223. *Corollary.* If we have

$$A = B$$

the semiaxis (363) becomes equal to A, so that the revolving hyperbola retains its original axes, and the surface is that of an hyperboloid of revolution. The sections made in this case, by a plane parallel to the plane of $x\,y$, are circles.

224. *Problem.* To investigate the form of the surface of equation (320).

Solution. By omitting the numbers below the letters, (320) may be written

$$\frac{x^2}{A^2} + \frac{y^2}{B^2} - \frac{z^2}{C^2} = 0. \qquad (364)$$

I. The section made by a plane parallel to the plane of xy, at the distance c, is an ellipse, of which the semiaxes are

$$\frac{A\,c}{C} \text{ and } \frac{B\,c}{C}; \qquad (365)$$

when the distance c is zero, this ellipse is a point.

II. The section, made by a plane parallel to the plane of of xz, at the distance c, is an hyperbola, of which the semitransverse axis, parallel to the axis of z, is

$$\frac{C\,b}{B} \text{ and the semiconjugate } \frac{A\,b}{B}. \qquad (366)$$

This section becomes the combination of the two straight lines

$$C\,x = \pm\, A\,z, \qquad (367)$$

when b is zero.

III. The section, made by a plane parallel to the plane of yz, at the distance a, is an hyperbola, of which the semitransverse axis, parallel to the axis of z, is

$$\frac{C\,a}{A} \text{ and the semiconjugate } \frac{B\,a}{A}. \qquad (368)$$

This section becomes the combination of the two straight lines

$$C\,y = \pm\, B\,z, \qquad (369)$$

when a is zero.

IV. The equation of the section, made by a plane passing through the axis of x, and inclined by an angle β' to the axis of y, is

$$\frac{x_1^2}{A^2} + \left(\frac{\cos^2 \beta'}{B^2} - \frac{\sin^2 \beta'}{C^2} \right) y_1^2 = 0. \qquad (370)$$

When the condition (356) is fulfilled, this section is reduced to the point

$$x_1 = 0, \; y_1 = 0.$$

But every section parallel to this one is an ellipse.

When the condition (358) is fulfilled, the section is reduced to the straight line

$$x_1 = 0 \, ;$$

that is, to the axis of y_1; and every section parallel to this is a parabola.

When the condition (359) is fulfilled, the section is the combination of the two straight lines

$$\frac{x_1}{A} = \pm \sqrt{\left(\frac{\sin.^2 \beta'}{C^2} - \frac{\cos.^2 \beta'}{B^2} \right)} \cdot y_1, \qquad (371)$$

and every section parallel to this is an hyperbola.

In the same way, and with like results, the sections made by a plane passing through the axis of y may be found.

V. The equation of the section, made by a plane passing through the axis of z, and inclined by an angle β to the axis of x, is by (329)

$$\left(\frac{\cos.^2 \beta}{A^2} + \frac{\sin.^2 \beta}{B^2} \right) y_1^2 - \frac{x_1^2}{C^2} = 0, \qquad (372)$$

so that this section is the combination of the two straight lines

$$\frac{x_1}{C} = \pm \sqrt{\left(\frac{\cos.^2 \beta}{A^2} + \frac{\sin.^2 \beta}{B^2} \right)} \cdot y_1, \qquad (373)$$

which are inclined at equal angles on opposite sides of the axis of x_1.

This surface may then be regarded as generated by a straight line which passes through the origin, and revolves about the axis of z, inclined to this axis by a variable angle, whose tangent is

$$\frac{AB}{C\sqrt{(B^2 \cos.^2 \beta + A^2 \sin.^2 \beta)}}; \qquad (374)$$

the surface is therefore that of a *cone*.

225. *Corollary.* If A and B are equal, the axes (366) are equal, and the section parallel to $x\,y$ is a circle; and the tangent (373) of the angle which the revolving lines makes with the axis of z, becomes

$$\frac{A}{C},$$

so that its value is constant, and the cone is *a right cone*.

226. All the curves of the second degree may then be obtained by cutting a right cone by different planes, these curves are therefore called *conic sections*.

From examining section IV. of art. 224, it appears that the section of a right cone is an *ellipse*, when the plane cuts completely across the cone, so as not to meet the cone produced above the vertex; it is a parabola, when the plane is parallel to one of the extreme sides of the cone, so as not to meet it, nor the cone produced above the vertex; it is an hyperbola, when the plane cuts the cone both above and below the vertex.

227. *Problem.* To investigate the form of the surface of equation (307).

14

Solution. By omitting the numbers below the letters, (307) becomes

$$B y^2 + C z^2 + M = 0. \tag{375}$$

The equation of the section made by a plane parallel to the plane of $y z$ is, then, the same with (375), so that the surface must be a cylinder, of which (375) is the equation of the base.

228. *Problem.* To investigate the form of the surface of equation (308).

Solution. By omitting the numbers below the letters, (308) becomes

$$B y^2 + C z^2 + H x = 0. \tag{376}$$

I. The section made by a plane parallel to the plane of $y z$ is an ellipse or an hyperbola, and those made by planes parallel to the planes of $x y$ and $x z$ are parabolas.

II. The equation of the section, made by a plane passing through the axis of x, and inclined by an angle β' to the axis of y, is

$$(B \cos.^2 \beta' + C \sin.^2 \beta') y_1^2 + H x_1 = 0, \tag{377}$$

so that it is a parabola, of which the vertex is the origin, and the parameter

$$2 p = - \frac{H}{2 (B \cos.^2 \beta' + C \sin.^2 \beta')}. \tag{378}$$

The surface may then be considered as generated by the revolution of a parabola, with a variable parameter, about the axis of x_1. It is, hence, called a *paraboloid.*

Cylinder.
---

229. *Corollary.* If B and C are equal, (378) becomes

$$2p = -\frac{H}{2B},\qquad (379)$$

so that the parameter is no longer variable, and the paraboloid is a paraboloid of revolution.

230. *Problem.* To investigate the form of the surface of equation (309).

Solution. By omitting the numbers below the letters, (309) becomes

$$Cz^2 + Hx + Iy = 0.\qquad (380)$$

Before proceeding to investigate the sections of the surface, we may refer it to other axes, which have the same origin, of which the axis of z_1 is the same with the axis of z, and the plane of $x_1 y_1$ the same with the plane of xy. In this case, we have

$$a = b = c = 0,$$

$$\alpha'' = \beta'' = \gamma = \gamma' = 90°,\ \gamma'' = 0$$

$$\beta' = \alpha,\ \beta = 90° + \alpha,\ \alpha' = \alpha - 90°,$$

so that (40, 41, 42) become

$$x = x_1 \cos.\,\alpha - y_1 \sin.\,\alpha \qquad (381)$$

$$y = x_1 \sin.\,\alpha + y_1^{\,1} \cos.\,\alpha \qquad (382)$$

$$z = z_1, \qquad (383)$$

which being substituted give, by taking α to satisfy the condition that the coefficient of y_1 is zero,

$$Cz_1^2 + (H\cos.\,\alpha + I\sin.\,\alpha)\,x_1 = 0, \qquad (384)$$

in which α is determined by the equation

$$\text{tang. } \alpha = \frac{I}{H}. \tag{385}$$

The equation of the section, which is made by a plane parallel to the plane of $x_1 \, z_1$, is now the same with (384); that is, the section is a parabola, and the surface is that of a cylinder, of which the base is a parabola.

231. *Problem,* To investigate the form of the surface of equation (310).

Solution. When (310) is possible, it is evidently the combination of the two equations

$$z_2 = \pm \sqrt{-\frac{M_1}{C_1}} \tag{386}$$

each of which represents a plane parallel to the plane of $x_2 \, y_2$.

DIFFERENTIAL CALCULUS.

BOOK II.

14*

BOOK II.

CHAPTER I.

FUNCTIONS.

1. A *variable* is a quantity, which may continually assume different values.

A *constant* is a quantity, which constantly retains the same value.

Thus the axes of an ellipse or hyperbola are constants, while the ordinates and abscissas are variables.

Constants are usually denoted by the first letters of the alphabet, and variables by the last letters, but this notation cannot always be retained.

2. When quantities are so connected together that changes in the values of some of them affect the values of the other, they are said to be *functions* of each other. Any quantity is, then, a function of all the quantities upon which its value depends; but it is usual to name only the variables of which it is a function.

Functions are denoted by the letters $f.$, $f.$, $F.$, $\varphi.$, $\psi.$, $\pi.$, $f.'$, $F.'$, $f_0.$, $f_1.$, &c. ; thus

$f.(x)$, $F.(x)$, $\varphi.(x)$, $\pi.(x)$, &c., $f.'(x)$, &c., $f_0.(x), f_1.(x), f_2.(x)$, &c. are functions of x.

$$f.(x,y),\ F.(x,y),\ \&c.$$

are functions of x and y.

3. When variables are functions of each other, some of them can always be selected, to which, if particular values are given, the corresponding values of all the rest can be determined.   The variables, which are thus selected, are called the *independent variables*.

4. When a function is actually expressed in terms of the quantities, upon which it depends, it is called an *explicit function*.

But when the relations only are given, upon which the function depends, the function is called an *implicit function*.

Thus the roots of an equation are, before its solution, implicit functions of its coefficients; but, after its solution, they are explicit functions.

5. A function of a variable may be expressed geometrically, by regarding it as the ordinate of a curve, of which the variable is the abscissa.

A function of two variables may be expressed geometrically, by regarding it as one of the coördinates of a surface, of which the two variables are the other two coördinates.

A function is said to be constructed, geometrically, when the curve or surface, which expresses it, is constructed.

The inspection of the curve or surface, which thus represents a function, is often of great assistance in obtaining a clear idea of the function.

6. *Algebraic functions* are those which are formed by addition, subtraction, multiplication, division, raising to given powers, whether integral or fractional, positive or negative.

An *integral function* is one, which contains only integral powers of the variable ; and a *rational fractional function* is a fraction, whose numerator and denominator are both integral functions.

Every other algebraic function is called *irrational.*

Thus

$$a + x, \ a - x, \ a x + b y, \ a + b x + c x^2 + \&c.$$

are integral functions ;

$$\frac{a}{x}, \ x^{-a}, \ \frac{a + b x + c x^2}{a' + b' x + c' x^2}$$

are rational fractional functions ;

and $\sqrt{x}, \ x^{\frac{3}{4}}, \&c.$

are irrational functions.

7. *Exponential or logarithmic functions* involve variable exponents or logarithms of variables.

Thus, a^x, log. x, &c. are logarithmic or exponential functions.

8. *Trigonometric or circular functions* involve trigonometric operations.

Thus   sin. x, tan. x, &c.
are trigonometric or circular functions.

9. *Compound functions* result from several successive operations.

Thus log. sin. x is the logarithm of the sine of x.

10. When functions are so related, that the compound function formed from their combination is independent of the order in which the functional operations
are performed, the functions are said to be *relatively
free;* otherwise they are *fixed.*

Thus if the two functions φ and f are so related that the
compound function $\varphi.f.x$ is equal to the compound function
$f.\varphi.x$, these two functions are *relatively free,* and this condition is algebraically

$$\varphi.f.x = f.\varphi.x ; \qquad (387)$$

or if we omit the variable, which is often done in functional
expressions which involve but one variable, (387) becomes

$$\varphi.f. = f.\varphi. \qquad (388)$$

11. A *linear* function is one, which leads to the same
result, whether the operation indicated by it is performed upon the whole of a polynomial at once, or upon the
different terms of it successively.

Such a function is indicated by the equation

$$f.(x \pm y) = f.x \pm f.y ; \qquad (389)$$

and the product $m\,x$, that is, m times the variable, is a simple
example of such a function.

12. *Theorem.* *The compounds of linear functions are linear.*

Proof. Let f and f' be two linear functions, we are to prove that

$$f.f.'(x \pm y) = f.f.'x \pm f.f.'y. \qquad (390)$$

Now we have from definition

$$f.'(x \pm y) = f.'x \pm f.'y, \qquad (391)$$

and therefore

$$.f.f.'(x \pm y) = f.(f.'x \pm f.'y) = f.f.'x \pm f.f.'y \qquad (392)$$

as we wished to prove.

13. When the same operation is successively repeated, the result is called *the second, third, &c. function.*

Thus the log. log. x is the second logarithm of x.

These repeated functions may be expressed by a notation similar to that of powers; thus

$$\log.^2 x = \log. \log. x$$

$$\log.^3 x = \log. \log.^2 x = \log. \log. \log. x, \&c.$$

$$f.^2 x = f.f.x$$

$$f.^3 x = f.f.^2 x = f.f.f. x, \&c.$$

Care must be taken not to confound $f^n(x)$ with $[f(x)]^n$, or with $f(x)^n$, which have widely different significations; thus, $[f(x)]^n$ is the nth power of the function of x, $f(x)^n$ is the function of the nth power of x, while $f^n(x)$ is the nth function of x.

The common use of a different notation in the case of trigonometric functions must, however, cause them to be excepted from these remarks ; thus, sin.$^n x$ and sin.x^n do either of them denote the nth power of sin.x. Whenever we extend this notation to trigonometric functions, we shall indicate it by enclosing the exponent within brackets; thus we shall denote the second, third, &c. sine of x by sin.$^{[2]} x$, sin.$^{[3]} x$, &c.

14. By a process of reasoning, precisely similar to that used in the case of powers, it may be proved that we must have

$$f.^m f.^n x = f.^{m+n} x; \qquad (393)$$

or, omitting the variable

$$f.^m f.^n = f.^{m+n}. \qquad (394)$$

This equation may be adopted as applicable to all functional exponents, whether positive or negative, entire or fractional; and the signification of the exponent, when not positive and integral, must, in this case, be determined by the aid of the equation.

15. *Problem.* To determine the signification of a function, of which *the exponent is zero.*

Solution. Equation (393) becomes by making

$$m = 0,$$

$$f.^0 f.^n x = f.^{0+n} x = f.^n x; \qquad (395)$$

so that if we put

$$f.^n x = y$$

we have $\qquad f.^0 y = y; \qquad (396)$

that is, *the function whose exponent is zero is the variable itself, and this function may be represented by unity.*

16. *Problem.* To determine the signification of a function, of which *the exponent is negative.*

Solution. Equation (393) becomes, by making

$$m = -n$$

$$f.^{-n} f.^{n} x = f.^{0} x = x; \qquad (397)$$

or if we put

$$f.^{n} x = y,$$

we have

$$f.^{-n} y = x; \qquad (398)$$

that is, *if two variables, x and y, are functions of each other, whatever function y may be of x, x is the corresponding negative function of y, or, as it is usually called, the inverse function of y.*

17. Confusion is likely to arise in the use of negative exponents, unless it is carefully observed that many functions have, like roots, several different values corresponding to the same value of the variable.

18. *Problem.* To determine the signification of a function, of which the *exponent is fractional.*

Solution. We have, from (394)

$$f.^{m} f.^{m} f.^{m} \ldots = (f.^{m})^{n} = f.^{mn}$$

in which n denotes the number of repetitions of $f.^{m}$. If now

$$n' = mn, \quad m = \frac{n'}{n}$$

$$f.^{mn} = f.^{n'}$$

$$f.^{m} = f.^{\frac{mn}{n}} = f.^{\frac{n'}{n}}$$

15

that is, if m is a fraction, of which the denominator is n and the numerator n', the corresponding function is one which, repeated n times, gives the n'th repetition of the original function f.

19. *Theorem*. When the different functions of which a compound function is composed are linear and relatively free, they may be combined precisely as if the letters which indicate them were factors instead of functional expressions.

Proof. For the two equations (388) and (389), which apply to this case, are the same in form as the two fundamental equations of addition and multiplication, upon which all arithmetical and algebraical processes are founded.

20. *Corollary*. The repetitions of the compound functions

$$f. + f_1., \ f. + f_1. + f_2., \ \&c.$$

in which f, f_1, f_2, &c. are linear and relatively free, may be effected by means of the binomial and polynomial theorems. Thus

$$(f. + f_1.)^n = f.^n + n f.^{n-1} f_1. + \frac{n.\,n-1}{1.2} f.^{n-2} f_1^2. + \&c. \quad (399)$$

$$(f. + f_1. + f_2.)^n = f.^n + n f.^{n-1} f_1. + n f.^{n-1} f_2. + \&c. \quad (400)$$

$$+ \frac{n(n-1)}{1.2} f.^{n-2} f_1^2. + \&c.$$

21. The exponents of functions are so similar to the exponents of powers, that they may be used in a similar way, and called *functional logarithms*; so that if any function, as $f.$, is assumed as a *base*, the functional log-

arithm of any other function indicates the exponent which the base must have to be equivalent to this new function.

Thus, if we denote functional logarithms by $[f.\log.]$, and

if $$f.^n = \varphi. \tag{401}$$

we have $$n = [f.\log.]\,\varphi. \tag{402}$$

and it may be shown, as in the theory of logarithms, that

$$[f.\log.]\,\varphi.\varphi.' = [f.\log.]\,\varphi. + [f.\log.]\,\varphi.' \tag{403}$$

$$[f.\log.]\,\varphi.^n = n\,[f\log.]\,\varphi. \tag{404}$$

22. When a function has but one finite value corresponding to each value of its variable, included between given limits, and varies by infinitely small degrees for infinitely small changes in the value of its variable between these limits, it is said to be *continuous* between these limits.

The curve which represents a continuous function is obviously a continuous curve.

CHAPTER II.

INFINITESIMALS.

23. *Theorem. Any power of an infinitesimal is infinitely smaller than any inferior power of the same infinitesimal.*

Demonstration. Let i be the given infinitesimal, a the exponent of the given power, and b the exponent of an inferior power. We have, then, the ratio

$$i^a : i^b = i^{a-b} : 1 ;$$

that is, i^a bears the same ratio to i^b that i^{a-b} does to unity, or i^a is infinitely smaller than i^b.

24. *Definition.* If a given infinitesimal is assumed as a *base* or standard to which all others may be referred, any infinitesimal is said to be of the *order a*, when it is infinitely less than any power of the base inferior to the ath power, and infinitely greater than any power of the base superior to the ath power.

25. *Corollary.* If A is a finite quantity, and i the infinitesimal which is assumed as the base, $A i^a$ must be an infinitesimal of the ath order.

26. *Corollary.* If, in the preceding article, we make

$$a = 0$$

Negative orders of infinitesimals.

we have
$$A\,i^{a} = A\,i^{0} = A\,;$$
so that *a finite quantity is an infinitesimal of the order zero.*

27. *Corollary.* If, in art. 12, we make a negative, or
$$a = -\,a'$$
we have
$$A\,i^{a} = A\,i^{-a'} = \frac{A}{i^{a'}} = \infty$$

so that *infinitely great quantities may be regarded as infinitesimals of negative orders;* that is, an infinitely great quantity of the ath order is an infinitesimal of the $-a^{\text{th}}$ order.

28. *Theorem. Of two infinitesimals of different orders, that, which is of the inferior order is the infinitely greater.*

Demonstration. Let I and J be the two infinitesimals of the orders a and b respectively, a being greater than b, and let c be any number between a and b, and let i be the base. We have, by the definition of art. 24, I infinitely less than i^{c}, and J infinitely greater than i^{c}, so that I is infinitely less than J, agreeably to the theorem to be demonstrated.

29. *Corollary. When infinitesimals of different orders are connected together by the signs of addition or subtraction, all may be neglected but those which are of the lowest order, so that the sum or difference is of the same order with those of the lowest order, which are retained.*

15*

30. *Corollary.* *The continued product of several in-finitesimals is of an order equal to the sum of the orders of the factors.*

31. *Corollary.* If one or more of the factors is finite, the product is of the order equal to the sum of the orders of the other factors.

32. *Corollary.* *The order of the quotient of one infinitely small quantity divided by another is equal to the order of the dividend diminished by that of the divisor.*

33. *Corollary.* The order of any power of an infinitesimal is equal to the order of the infinitesimal multiplied by the exponent of the power; and the order of any root is equal to that of the infinitesimal divided by the exponent of the root.

. 34. *Theorem.* *The order of any function of an infinitesimal is equal to the product of the order of the infinitesimal multiplied by the order which the function would have if the infinitesimal were assumed as the base.*

Proof. Let I be the infinitesimal of the order a, J a function of I, which would be of the order b, if I were the base, and let i be the base; we are to prove that J is of the order ab.

Since J is of the order b with reference to I as a base, it is, by art. 24, infinitely greater than I^c when c is greater than b, and infinitely less than I^c when c is less than b; but I^c is, by art. 33, of the order ac, or of the same order with i^{ac}; and, therefore, J is infinitely less than i^{ac} when ac is less than ab, and infinitely greater than i^{ac} when ac is greater than ab; or J is, by art. 24, of the order ab.

35. Corollary. The order of any function of an infinitesimal of the first order is of the same order with the same function of the base.

36. Corollary. The ratios between the orders of several infinitesimals are not changed by changing the base; and *their orders with reference to the new base are obtained by dividing their original orders by the order of the new base referred to the original one.*

This rule cannot, however, be applied when the order of the new base, referred to the original one, is zero or infinity.

37. Corollary. If I is an infinitesimal of the order a with reference to the base i, i must be of the order $\dfrac{1}{a}$ with reference to I as a base.

38. Problem. To find the order of a^i, when i is the infinitesimal base.

Solution. Denote the order of a^i by x; that of $(a^i)^m$ is, by art. 33, mx; while that of a^{mi} is, by art. 35, x; but we have

$$(a^i)^m = a^{mi}$$

and, therefore,

$$mx = x,$$

which gives

$$x = 0;$$

that is, the order of a^i is zero.

39. *Problem.* To find the order of log. i, when i is the base.

Solution. Denote the order of log. i by x; that of a log. i is, by art. 31, also x; while that of log. i^a is, by art. 34, $a\,x$; but we have

$$a \log. i = \log. i^a,$$

and, therefore, $a\,x = x,$

which gives $x = 0$;

that is, the order of log. i is zero.

40. *Corollary* The order of $\dfrac{1}{\log. i}$ is also zero.

41. If i were zero, we know that a^i, log. i, and $\dfrac{1}{\log. i}$ would be respectively 1, $-\infty$ and 0; and their values must differ infinitely little from these values, when i is infinitely small; that is, a^i is finite, log. i is infinite, and $\dfrac{1}{\log. i}$ is infinitely small. But each of these infinitesimals has been shown to be of the order zero, so that there are infinitesimals of the order zero, which are finite, infinite, and infinitely small.

42. *Problem.* To find the nth power of $1 + i$, when i is infinitely small and n infinitely great, so that

$$n\,i = a. \tag{405}$$

Solution. The binomial theorem gives

$$(1+i)^n = 1 + n\,i + \frac{n\,(n-1)}{1.2}\,i^2 + \frac{n(n-1)(n-2)}{1.2.3}\,i^3 + \&\text{c.} \tag{406}$$

But n is infinite, and, therefore,

$$n - 1 = n, \ n - 2 = n, \ \&\text{c.} \tag{407}$$

which substituted in (406) give, by (405),

$$(1+i)^n = 1 + ni + \frac{n^2 i^2}{1.2} + \frac{n^3 i^3}{1.2.3} + \&c.$$

$$= 1 + a + \frac{a^2}{1.2} + \frac{a^3}{1.2.3} + \&c. \qquad (408)$$

43. Corollary. When

$$a = 1$$

(408) becomes

$$(1+i)^{\frac{1}{i}} = 1 + 1 + \frac{1}{1.2} + \frac{1}{1.2.3} + \&c. \qquad (409)$$

If we denote the value of the second member of (409) by e, we easily find

$$e = 2{,}71828 + \qquad (410)$$

and (409) is

$$(1+i)^{\frac{1}{i}} = e \qquad (411)$$

44. Corollary. Since

$$n = \frac{a}{i}$$

we have

$$(1+i)^{\frac{a}{i}} = \left((1+i)^{\frac{1}{i}} \right)^a = e^a = 1 + a + \frac{a^2}{1.2} + \&c. \quad (412)$$

45. The number e is the base of Neper's system of logarithms, and the logarithms taken in this system are called the *Neperian logarithms*. The Neperian logarithms will be generally used in the course of this work, and will be denoted by log. as usual.

Exponential infinitesimal functions.

---

46. *Corollary.* The log. of (411) is

$$\frac{1}{i} \log. (1 + i) = \log. e = 1 \tag{413}$$

$$\log. (1 + i) = i \tag{414}$$

47. *Corollary.* If in (412) we put for a the value

$$a = m\,i \tag{415}$$

and transpose 1 to the first member, (412) becomes

$$e^{m\,i} - 1 = m\,i; \tag{416}$$

all the terms of the second member except the first being omitted, because they are infinitely smaller.

48. *Corollary.* If b is taken so that

$$b = e^m \text{ or } m = \log. b, \tag{417}$$

(416) becomes

$$b^i - 1 = i \log. b. \tag{418}$$

49. *Corollary.* If in (416)

$$m = 1, \tag{419}$$

(416) is

$$e^i - 1 = i. \tag{420}$$

CHAPTER III.

DIFFERENTIALS.

50. The *difference* of a function is the difference between its two values, which correspond to two different values of the variable.

When the difference between the two values of the variable is infinitely small, the difference of the function is called its *differential*.

The letters $\varDelta$., $\varDelta$.', $\varDelta$.", &c., $\varDelta$.$_0$, $\varDelta$.$_1$, &c., ν., ν.', &c., placed before a function denote its differences, and corresponding differences are denoted by the same letters. Thus

$$\varDelta.f.x, \quad \varDelta.f.'x, \quad \&c.$$

are corresponding differences of $f.(x)$, $f.'(x)$, &c., and these differences correspond to the difference $\varDelta.x$ of the variable, so that

$$\varDelta.f.x = f.(x + \varDelta.x) - f.x. \qquad (421)$$

$$\varDelta.f.'x = f.'(x + \varDelta.x) - f.'x. \ \&c.$$

Differentials are denoted by the letters δ, δ', &c., d, d', &c,. D, D', &c.

51. *Theorem. Differences and differentials are linear functions.*

Proof. For it is obvious that the increment of the sum of several functions arising from an increase of the variable is equal to the sum of the increments of the variable ; or, as it may be expressed algebraically, that

$$\varDelta.(f.\pm f.') = [(f. + \varDelta.f.) \pm (f.' + \varDelta.f.')] - (f.\pm f.')$$
$$= \varDelta.f. \pm \varDelta.f.' \tag{422}$$

52. *Theorem. Differences and differentials are free relatively to any other linear function.*

Proof. Let $f.$ be a linear function, and (389) gives

$$f.(x + \varDelta. x) = f. x + f. \varDelta. x, \tag{423}$$

which, substituted in (421), gives

$$\varDelta.f. x = f. \varDelta. x, \tag{424}$$

and this is the theorem to be proved.

53. *Corollary.* In equation (399) we may put

$$f_1. = \varDelta.$$

and it becomes

$$(f. + \varDelta.)^n = f.^n + n f.^{n-1} \varDelta. + \&c. \tag{425}$$

54. *Corollary.* If the function $f.$ of (425) is unity, (425) becomes

$$(1. + \varDelta.)^n = 1. + n \varDelta. + \frac{n(n-1)}{1.2} \varDelta.^2 + \&c. \tag{426}$$

55. In finding differences and differentials, the differences and differentials of the independent variables are also independent and may vary or not, as may be

most convenient. It is usually most simple to suppose the differences and differentials of the independent variables not to vary, and, adopting this hypothesis, we must *regard the second differences and differentials of the independent variables as constant.*

56. *Corollary.* By the principle of art. 55, (426) becomes when applied to the independent variable x

$$(1. + \varDelta.)^n x = (1. + n \varDelta.) x = x + n \varDelta. x \qquad (427)$$

for we have

$$\varDelta.^2 x = 0, \qquad \varDelta.^3 x = 0, \text{ \&c.} \qquad (428)$$

57. *Corollary.* Taking the function $f.$ of each member of (427), we have

$$f. (1. + \varDelta.)^n x = f. (1. + n \varDelta.) x. \qquad (429)$$

58. *Corollary.* Equation (423) gives, by transposition and omitting x,

$$f. + \varDelta. f. = f. (1. + \varDelta.) \qquad (430)$$

or $\qquad (1. + \varDelta.) f. = f. (1. + \varDelta.) \qquad (431)$

so that the functions $f.$ and $(1. + \varDelta.)$ are relatively free, and we have, by (429),

$$(1. + \varDelta.)^n f. x = f. (1. + \varDelta)^n x = f. (1. + n \varDelta.) x$$
$$= f. (x + n \varDelta. x) \qquad (432)$$

59. *Corollary.* Since the linear function $\varDelta.$ may be subjected to all the forms of algebraic calculation, it may be substituted for a in (412), and gives

16

Differential coefficient.

$$e^{\varDelta \cdot} = \left(1. + \varDelta. + \frac{\varDelta.^2}{1.2} + \&c. \right) ;$$ (433)

and in like manner

$$e^{n \varDelta \cdot} = \left(1. + n\varDelta. + \frac{n^2 \varDelta.^2}{1.2} + \&c. \right).$$ (434)

60. The quotient of the differential of a function divided by the differential of the variable is called the *differential coefficient* of the function ; the differential coefficient of the differential coefficient is the *second differential coefficient*, and so on.

Differential coefficients are denoted by d_c. ; thus

$$d_c. f. x = \frac{d f. x}{d x}$$ (435)

$$d_c^2. f. x = d_c. d_c. f. x ;$$ (436)

or, since $d x$ is independent of x,

$$d_c^2. f. x = \frac{d. d_c. f. x}{d x} = \frac{d^2 f. x}{d x^2}.$$ (437)

61. *Theorem.* The differential coefficients of continuous functions are finite functions of the variable, independent of the differential of the variable.

Proof. I. Let BC (fig. 47.) be the curve which denotes the function $f. x$, so that if A is the origin, we have for

$$AP = x, \quad PM = f. x,$$

and if we take

$$PP' = d. x = MR, \quad PP'' = d'. x = MR',$$

we have

$$d.f.x = P'M' - PM = M'R,$$

$$d'.f.x = P''M'' - PM = M''R',$$

$$d_c.f.x = \frac{M'R}{MR}, \quad d'_c.f.x = \frac{M''R'}{MR'}.$$

But since $MM'M''$ is an infinitely small portion of the curve, it may be regarded as a straight line, and we have

$$\frac{M'R}{MR} = \frac{M''R'}{MR'},$$

or $\qquad d_c.f.x = d'_c.f.x\,;$

that is, the value of the differential coefficient does not change with that of the differential.

II. The differential coefficient is, in general, a finite function, for the ratio $M'R : MR$, which represents this function, is the tangent of the angle $M'MR$, by which the curve is inclined to the axis AX.

62. *Corollary.* We have, by (435) and (437),

$$d.f.x = d_c.f.x.dx, \quad d^2.f.x = d_c^2.f.x.dx^2, \qquad (438)$$

so that if dx is an infinitesimal of the first order, $df.x$ is, by art. 30, of the same order, $d^2f.x$ is of the second order, and so on.

Differentials may then be regarded as infinitesimals of the same order with their exponents.

63. *Corollary.* If we put

$$n\,dx = h, \text{ so that } n = \frac{h}{dx}, \qquad (439)$$

and put $d.$ for $\varDelta.$ in (432), we have

$$(1. + d.)^n f. x = f. (x + h);\qquad(440)$$

or developing, as in art. 42, and putting

$$d_c. = \frac{d.}{dx}, \quad d_c^2. = \frac{d^2.}{dx^2}\qquad(441)$$

$$\left(1. + h\, d_c. + \frac{h^2\, d_c^2.}{1.2} + \&c.\right) f.x = f. (x + h).\quad(442)$$

But, by (434),

$$1. + h\, d_c. + \frac{h^2\, d_c^2}{1.2} + \&c. = e^{h.\, d_c.}\qquad(443)$$

whence

$$e^{h\, d_c} \cdot f.\, x = f. (x + h).\qquad(444)$$

64. *Corollary.* When, in (442) and (444), we put

$$x = 0$$

they become

$$\left(1. + h\, d_c. + \frac{h^2\, d_c^2.}{1.2} + \&c.\right) f.0 = f.h\qquad(445)$$

$$e^{h\, d_c} f.\, 0 = f.\, h\qquad(446)$$

and if we now put x for h, we have

$$\left(1 + x\, d_c + \frac{x^2\, d_c^2}{1.2} + \&c.\right) f.\, 0 = f.x\qquad(447)$$

$$e^{x\, d_c.} f.\, 0 = f.x.\qquad(448)$$

Equation (442) is called *Taylor's theorem,* of which (444) is a neat form of writing, and (447) is called *Mac Laurin's theorem.* The great use of these theorems will be seen in the sequel.

Increasing or decreasing function.
---

65. *Corollary.* If we put, in (442) and (444), .

$$h = \mathit{\Delta}.x,$$

they become, by (431),

$$(1+\mathit{\Delta}.x\,d_c. +\&c.)f.x = f.(1.+\mathit{\Delta}.)x = (1.+\mathit{\Delta}.)f.x \quad (449)$$

$$e^{\mathit{\Delta}.x\,d_c.}f.x = (1.+\mathit{\Delta}.)f.x. \quad (450)$$

66. *Theorem.* When the differential coefficient of a continuous function is finite and positive, the function increases with the increase of the variable; but if the differential coefficient is negative, the function decreases with the increase of the variable.

Proof. For the differential coefficient is the ratio between the differential of the function and that of the variable, and is therefore positive when both these differentials are positive, and negative when one is positive and the other negative.

67. *Corollary.* If the variable increases from any of its values, for which the function vanishes, the function must be positive if the differential coefficient is positive, and negative if the differential coefficient is negative; that is, the function has the same sign with the differential coefficient. The reverse is the case if the variable decreases.

68. *Theorem. The greatest value of the differential coefficient of a continuous function, which vanishes with the variable, and extends to a given limit, is larger than the quotient of the greatest value of the function, divided by the corresponding value of the variable; and*

16*

the smallest value of the differential coefficient is smaller than the smallest value of this quotient.

Proof. Let $f.$ denote the given function which vanishes with the variable, let x' be the limit of the variable to which the function is extended, and let A and B be respectively the greatest and the least of the values of the differential coefficient, so that

$$A - d_c f. x \text{ and } d_c f. x - B$$

are positive. But these two quantities are the differential coefficients of the functions

$$A x - f. x \text{ and } \bar{f}. x - B x,$$

both of which vanish when x is zero, and are, therefore, by art. 67, of the same sign with their differential coefficients when x is increasing and positive, and of the opposite sign when x is decreasing and negative. Hence these two functions are of the same sign with x, and their quotients, divided by x, must be positive, that is,

$$A - \frac{f. x}{x} \text{ and } \frac{f. x}{x} - B$$

are positive. It follows, then, that A is greater than the quotient of $f. x$ divided by x, and that B is less than this quotient.

The truth of this proposition may be exhibited geometrically. Thus, if AMB (fig. 48.) is the curve which represents $f. x$, so that, for

$$AP = x, \text{ we have } MP = f. x;$$

if the curve at M is produced in the straight line MT, we have, by art. 61,

$$d_e. f. x = \text{tang. } MTX,$$

and, by joining AM,

$$\text{tang. } MAX = \frac{MP}{AP} = \frac{f.x}{x}. \tag{451}$$

Now, M_1AX being the greatest value which the angle MAX has in this curve, it is evident that in proceeding from M_1 to A the curve must be inclined to the axis AX by an angle greater than M_1AX; so that at M, for instance, the value of tang. MTX, or of $d_e f.x$, is greater than tang. M_1AX, the greatest value of $\frac{f.x}{x}$; and, therefore, A, the greatest value of $d_e f. x$, must be greater than any value of $\frac{f.x}{x}$.

Again, M_2AX being the least value of MAX, we see that in proceeding from M_2 to A the curve must be inclined to the axis AX by an angle less than M_2AX, so that at M', for instance, we have

$$\text{tang. } M'T'X < \text{tang. } M_2AX;$$

whence it follows as above, that B is less than any value of

$$\frac{f.x}{x}.$$

69. *Theorem.* If a function f. and its differential coefficient are continuous, and if the function vanishes when the variable is zero, there is, for every value of the variable x, a value of ε less than unity, which satisfies the equation

$$f.x = x\, d_e f. (\varepsilon x). \tag{452}$$

Proof. If A and B are respectively the greatest and least values of $d_e f. x$, contained between the limits

$$x = 0, \text{ and } x = x,$$

it follows from the fact, that $d_c f. x$ is continuous, that it must assume every possible value between A and B, while x varies from 0 to x. But the value of $\dfrac{f. x}{x}$ is included between A and B, and, therefore, there is a value of x' less than x such that

$$d_c f. x' = \frac{f. x}{x};$$

or $\qquad\qquad f. x = x\, d_c f. x'. \qquad\qquad (453)$

But since x' is less than x, if we put

$$\theta = \frac{x'}{x} \text{ or } x' = \theta x,$$

we have θ less than unity, and (453) becomes (452).

70. *Corollary.* If in (452) divided by x, we suppose x to be such a function f_1 of a new variable x_1 as constantly to increase with the increase of x_1 and to vanish with x_1, so that

$$x = f_1. x_1,$$

and take θ_1, so that

$$\theta x = f_1. (\theta_1 x_1),$$

then θ_1 is evidently less than unity, and if we suppose

$$F. = f. f_1.$$

(452) becomes, by dividing by x,

$$\frac{f. x}{x} = \frac{d f. (\theta x)}{d (\theta x)} = \frac{f. f_1. x_1}{f_1. x_1} = \frac{d f. f_1. (\theta_1 x_1)}{d f_1. (\theta_1 x_1)},$$

or

$$\frac{F. x_1}{f_1. x_1} = \frac{d F. (\theta_1 x_1)}{d f_1. (\theta_1 x_1)} = \frac{d_c F. (\theta_1 x_1)}{d_c f_1. (\theta_1 x_1)};$$

Ratio of functions.

or, omitting the numbers below the letters, which are no longer needed, and put $x' = \theta x$

$$\frac{F.x}{f.x} = \frac{d_c F.(\theta x)}{d_c f.(\theta x)} = \frac{d_c F.x'}{d_c f.x'}. \tag{454}$$

71. *Corollary.* If the n first successive differential co-efficients of the functions $F.$ and $f.$ were continuous, and all but the nth vanished, (454) would give

$$\frac{F.x}{f.x} = \frac{d_c F.x'}{d_c f.x'} = \frac{d_c^2 F.x''}{d_c^2 f.x''} = \frac{d_c^n F.x_n}{d_c^n f.x_n}, \tag{455}$$

in which x, x', x'', &c. are decreasing, so that if we put

$$\theta_n = \frac{x_n}{x}$$

we have

$$\frac{F.x}{f.x} = \frac{d_c^n F.(\theta_n x)}{d_c^n f.(\theta_n x)}, \tag{456}$$

in which θ is less than unity.

CHAPTER IV.

COMPOUND AND ALGEBRAIC FUNCTIONS.

72. Theorem. The differential of a compound function of several simple functions, is equal to the sum of its partial differentials arising from allowing each simple function to vary by itself, independently of the other simple functions.

Proof. Let $f., f.'$ be simple functions, of which $\varphi.$ is the compound function. We will denote by $d_f. \; d_{f.'}$ the partial differentials, supposing $f., f'$ respectively to vary by themselves. We are to prove that

$$d\varphi.(f.,f.') = d_{f.} \, \varphi. \, (f.,f.') + d_{f.''} \, \varphi. \, (f.,f.') \qquad (457)$$

Now we have, by definition,

$$d_{f.} \, \varphi. \, (f.,f.') = \varphi. \, (f. + df., f.') - \varphi. \, (f.,f.'), \qquad (458)$$

or, by transposition,

$$\varphi. \, (f.,f.') = \varphi. \, (f. + df., f.') - d_{f.} \, \varphi. \, (f.,f.')$$

The differential of this last equation, supposing $f.'$ to vary, is

$$d_{f.'} \, \varphi.(f.,f.') = d_{f.'} \, \varphi \, (f. + df., f.') - d^2_{f.,f.'} \, \varphi. \, (f.,f.'). \qquad (459)$$

But, by definition,

$$d_{f.'} \varphi.(f. + df., f.') = \varphi.(f. + df., f.' + df.') - \varphi.(f. + df., f.') \qquad (460)$$

Differential of product.

which, substituted in (459), gives

$$d_{f.} \, \varphi. \, (f.,f.') = \varphi. \, (f. + df.,f.' + df.')$$
$$- \varphi. \, (f. + df.,f.') - d.^2 f.,f.' \, \varphi. \, (f.,f.'). \quad (461)$$

The sum of (458) and (461) is

$$d_{f.} \, \varphi. \, (f.,f.') + d_{f.'} \, \varphi. \, (f.,f.') = \varphi. (f.+df.,f.'+df.')$$
$$- \varphi. \, (f.,f.') - d^2_{f.,f'} \, \varphi. \, (f.,f.'). \quad (462)$$

But we have

$$d \, \varphi. \, (f.,f.') = \varphi. \, (f. + df.,f.' + df.') - \varphi. \, (f.,f.') \quad (463)$$

which, substituted in (462), gives (457) by omitting the last term, because it is a second differential, and therefore an infinitesimal of the second order.

73. Corollary. Equation (457) is written by omitting the function,

$$d. = d_{f.} + d_{f.}. \quad (464)$$

74. Problem. To differentiate $a \, f. \, x$.

Solution. We have, by definition,

$$d.(a f. x) = a f. (x + d x) - a f. x$$
$$= a[f.(x+ d x) - f. x] = a \, d.f.x. \quad (465)$$

75. Corollary. We have then

$$d. (a x) = a \, d x. \quad (466)$$

76. Corollary. We have also

$$d_{f.} \, (f. x . f.' x) = f.' \, x \, d f. x$$
$$d_{f'.} \, (f. x . f.' x) = f. \, x \, d f.' x$$

and, by (464),

$$d.(f.x.f.'x) = f.'x\,df.x + f.x\,df.'x. \qquad (467)$$

In the same way, if u and v are functions of x,

$$d.(uv) = u\,dv + v\,du. \qquad (468)$$

77. *Corollary.* Equation (467), divided by $f.x.f.'x$, is

$$\frac{d.(f.x.f.'x)}{f.x.f.'x} = \frac{d.f.x}{f.x} + \frac{d.f.'x}{f.'x}. \qquad (469)$$

78. *Corollary.* From (469), it follows that

$$\frac{d.(f.x.f.'x.f.''x\ldots)}{f.x.f.'x.f.''x\ldots} = \frac{d.f.x}{f.x} + \frac{d.f.'x}{f.'x} + \&\text{c}. \qquad (470)$$

79. *Corollary.* If in (470) we have the n functions

$$f.x = f.'x = f.''x = \&\text{c}. \qquad (471)$$

(470) becomes

$$\frac{d.(f.x)^n}{(f.x)^n} = \frac{n\,df.x}{f.x} \qquad (472)$$

which, freed from fractions, is

$$d\,(f.x)^n = n\,(f.x)^{n-1}\,df.x. \qquad (473)$$

Hence, *to differentiate any power of a function, multiply by the exponent and by the differential of the function, and diminish the exponent by unity.*

80. *Scholium.* The proof of (472) and (473), which is given in art. 79, is limited to positive integral exponents, but may be easily extended.

Differential of power.

I. *Proof of* (472) and (473) *for fractional exponents.* Let n be a fraction

$$n = \frac{m}{m'}$$

and let $\qquad \varphi.x = (f.x)^n$

so that $\qquad (\varphi.x)^{m'} = (f.x)^m.$

Equation (472) gives

$$\frac{m'\, d\, (\varphi.x)}{\varphi.x} = \frac{m\, d\, (f.x)}{f.x}$$

or

$$\frac{d\,(\varphi.x)}{\varphi.x} = \frac{d\,(f.x)^n}{(f.x)^n} = \frac{m}{m'}. \quad \frac{d\,(f.x)}{f.x} = \frac{n\,d\,(f.x)}{f.x}$$

which includes (472), and consequently (473).

II. *Proof of* (472) and (473) *for negative exponents.* Let n be negative

$$n = -m$$

and let $\qquad \varphi.x = (f.x)^{-m} = \frac{1}{(f.x)^m}$

so that $\qquad \varphi.x\,(f.x)^m = 1.$

The differential of which gives, by (470) and (472),

$$\frac{d\,\varphi.x}{\varphi.x} + \frac{d\,(f.x)^m}{(f.x)^m} = \frac{d\,\varphi.x}{\varphi.x} - \frac{m\,d\,f.x}{f.x} = 0$$

whence

$$\frac{d\,(f.x)^n}{(f.x)^n} = -\frac{m\,d\,f.x}{f.x} = \frac{n\,d\,f.x}{f.x}$$

which is the same as (472), and therefore includes (473).

17

Binomial theorem.

---

81. *Corollary.* Equation (483) gives

$$d \cdot x^m = m \, x^{m-1} \, d \, x \tag{474}$$

and $\quad d_c \cdot x^m = m \, x^{m-1} \tag{475}$

$$d_{.c}^2 \cdot x^m = m \, d_c \cdot x^{m-1} = m \, (m-1) \, x^{m-2} \tag{476}$$

$$d_c^3 \cdot x^m = m \, (m-1) \, (m-2) \, x^{m-3}, \ \&c. \tag{477}$$

If now we substitute x^m for $f \cdot x$ in (442), we have

$$(x+h)^m = x^m + h \, d_c \cdot x^m + \frac{h^2}{1 \cdot 2} \, d_c^2 \cdot x^m + \&c.$$

$$= x^m + m \, x^{m-1} h + \frac{m(m-1)}{1 \cdot 2} x^{m-2} h^2 + \&c. \tag{478}$$

which is the *binomial theorem.*

82. EXAMPLES FOR DIFFERENTIATION.

1. Differentiate $a \, x^m + b$.

    Ans. The differential coefficient is $m \, a \, x^{m-1}$.

2. Differentiate $\sqrt{(f \cdot x)}$.   *Ans.* $\dfrac{d f \cdot x}{2 \sqrt{(f \cdot x)}}$.

3. Differentiate $\dfrac{1}{f \cdot x}$.   *Ans.* $- \dfrac{d f \cdot x}{(f \cdot x)^2}$.

4. Differentiate $\dfrac{a}{(f \cdot x)^n}$.   *Ans.* $- \dfrac{n \, a \, d f \cdot x}{(f \cdot x)^{n+1}}$.

5. Differentiate $\dfrac{f \cdot x}{f' x}$.   *Ans.* $\dfrac{f' x \cdot d f \cdot x - f \cdot x \cdot d f' x}{(f' x)^2}$.

6. Differentiate $\dfrac{(f \cdot x)^m}{(f' x)^{m'}}$.

    Ans. $\dfrac{(f \cdot x)^{m-1} \, (m \, f' x \cdot d f \cdot x - m' \, f \cdot x \, d f' x)}{(f' x)^{m'+1}}$.

Differential of polynomial.

7. Find the successive differential coefficients of

$$a + b x + c x^2 + \ldots \ldots + M x^m.$$

Ans.   The first is $b + 2 c x \ldots + m M x^{m-1}$;

the second is $2 c \ldots + m (m-1) M x^{m-2}$;

the mth is $m (m-1) (m-2) \ldots 1 . M$;

the $(m+1)^{\text{st}}$ is $0.$

CHAPTER V.

LOGARITHMIC FUNCTIONS.

83. *Problem.* To differentiate a^x.

Solution. We have, by definition,

$$d\, a^x = a^{x+dx} - a^x = a^x\, a^{dx} - a^x$$
$$= a^x\, (a^{dx} - 1). \qquad \qquad \cdot \quad (479)$$

But, by (418),

$$a^{dx} - 1) = \log.\, a.\, dx\,; \qquad (480)$$

whence

$$d\, a^x = \log.\, a.\, a^x\, dx. \qquad (481)$$

84. *Corollary.* Hence

$$d_c\, a^x = \log.\, a.\, a^x$$
$$d_c^2.\, a^x = \log.\, a.\, d_c\, a^x = (\log.\, a)^2\, a^x$$
$$d_c^n.\, a^x = (\log.\, a)^n\, a^x \qquad (482)$$

and by making $\quad x = 0$

$$a^0 = 1$$
$$d_c.\, a^0 = \log.\, a$$
$$d_c^2.\, a^0 = (\log.\, a)^2$$
$$d_c^n.\, a^0 = (\log.\, a)^n. \qquad (483)$$

If now we put in (447)

$$f. \, x = a^x$$

we have

$$a^x = 1 + \log. \, a. \, x + \frac{(\log. \, a. \, x)^2}{1 \cdot 2} + \&c. \qquad (484)$$

85. *Corollary.* If we have $a = e$, we have

$$\log. \, e = 1,$$

$$e^x = d_c. \, e^x = d_c^2. \, e^x = d_c^n. \, e^x \qquad (485)$$

$$1 = d_c. \, e^0 = d_c^2. \, e^0 = d_c^n. \, e^0 \qquad (486)$$

$$e^x = 1 + x + \frac{x^2}{1 \cdot 2} + \&c. \qquad (487)$$

and (487) is the same with (412).

86. *Problem.* To differentiate log. x.

Solution. We have, by definition,

$$d. \log. \, x = \log. \, (x + d \, x) - \log. \, x$$

$$= \log. \frac{x + d \, x}{x} = \log. \left(1 + \frac{d \, x}{x} \right). \qquad (488)$$

But, by (414),

$$\log. \left(1 + \frac{d \, x}{x} \right) = \frac{d \, x}{x}$$

and, therefore,

$$d. \log. \, x = \frac{d \, x}{x}. \qquad (489)$$

17*

---

87. *Corollary.* Hence

$$d_c . \log . x = \frac{1}{x} \tag{490}$$

$$d_c^2 . \log . x = d_c . \frac{1}{x} = -\frac{1}{x^2}$$

$$d_c^3 . \log . x = \frac{2}{x^3}$$

$$d_c^4 . \log . x = -\frac{2.3}{x^4}$$

$$d_c^n . \log . x = \pm \frac{1.2. \ldots (n-1)}{x^n}, \tag{491}$$

the upper sign being used when n is odd, and the lower when n is even.

If now in (442) we make

$$f. = \log .$$

we have,

$$\log . (x + h) = \log . x + \frac{h}{x} - \frac{h^2}{2 x^2} + \frac{h^3}{3 x^3} - \&c. \tag{492}$$

88. *Corollary.* If in (492) we put

$$x = 1,$$

we have

$$\log . (1 + h) = h - \tfrac{1}{2} h^2 + \tfrac{1}{3} h^3 - \tfrac{1}{4} h^4 + \&c. \tag{493}$$

89. EXAMPLES.

1. Differentiate $\log . [x + \sqrt{(1 + x^2)}]$.

$$Ans \quad \frac{d x}{\sqrt{(1 + x^2)}}.$$

2. Differentiate $(\log . x)^n$.

$$Ans \quad \frac{n (\log . x)^{n-1} d x}{x}.$$

Differential of 2d and 3d logarithms.

3. Differentiate $\log^2 x$. $Ans.$ $\dfrac{d x}{x \log x}$.

4. Differentiate $\log^3 x$. $Ans.$ $\dfrac{d x}{x \log x \cdot \log^2 x}$.

5. Differentiate a^{b^x}.

   $Solution.$ Let $y = b^x$

   and we have $a^{b^x} = a^y$

   whence $d. a^{b^x} = d. a^y = \log a \cdot a^y \cdot d y$

   But $d y = d. b^x = \log b \cdot b^x \cdot d x$

   so that $d. a^{b^x} = \log a \cdot \log b \cdot a^{b^x} \cdot b^x \, d x$.

6. Differentiate x^y.

   $Solution.$ Equation (464) gives

   $$d. x^y = d_x. x^y + d_y. x^y$$

   But by (473) and (481)

   $$d_x. x^y = y x^{y-1} \, d x$$

   $$d_y. x^y = \log x \cdot x^y \, d y$$

   so that

   $$d. x^y = y x^{y-1} \, d x + \log x \cdot x^y \, d y.$$

Differential of sine and cosine.

CHAPTER VI.

CIRCULAR FUNCTIONS.

90. *Problem.* To differentiate sin. x.

Solution. We have, by definition,

$$d. \sin. x = \sin. (x + dx) - \sin. x.$$

But, by trigonometry,

$$\sin. (x + dx) = \sin. x \cos. dx + \cos. x \sin. dx,$$

$$\cos. dx = 1, \quad \sin. dx = dx,$$

so that $\qquad d. \sin x = \cos. x. dx.$ (494)

91. *Problem.* To differentiate cos. x.

Solution. Substitute in (494), $\frac{1}{2}\pi - x$, and it becomes

$$d. \sin. (\tfrac{1}{2}\pi - x) = \cos. (\tfrac{1}{2}\pi - x). d (\tfrac{1}{2}\pi - x). \quad (495)$$

But we have

$$\sin. (\tfrac{1}{2}\pi - x) = \cos. x, \quad \cos. (\tfrac{1}{2}\pi - x) = \sin. x$$

$$d (\tfrac{1}{2}\pi - x) = - dx,$$

which, substituted in (495), give

$$d. \cos. x = - \sin. x \, dx. \qquad (496)$$

Development of sine and cosine.

92. Corollary. Equations (494) and (496) give

$$d_c. \sin. x = \cos. x$$

$$d_c^2. \sin. x = d_c. \cos. x = -\sin. x$$

$$d_c^3. \sin. x = d_c^2. \cos. x = -d_c \sin. x = -\cos. x$$

$$d_c^4. \sin. x = d_c^3. \cos. x = -d_c \cos. x = \sin. x$$

$$d_c^n. \sin. x = d_c^{n-1}. \cos. x = d_c^{n-4}. \sin. x, \qquad (497)$$

so that the four values of all the successive differential coefficients of sin. x and cos. x are alternately cos. x, — sin. x, — cos. x, and sin. x.

Hence, making $x = 0$,

we have, when n is even,

$$d_c^n. \sin. 0 = d_c^{n-1} \cos. 0 = 0, \qquad (498)$$

but when n is odd

$$d_c^n. \sin. 0 = d_c^{n-1} \cos. 0 = \pm 1; \qquad (499)$$

the upper sign being used when $n - 1$ is divisible by 4, and the lower sign when $n + 1$ is divisible by 4.

These values, substituted in (447), give

$$\sin. x = x - \frac{x^3}{1.2.3} + \frac{x^5}{1.2.3.4.5} + \frac{x^7}{1.2.3.4.5.6.7} + \&c. \quad (500)$$

$$\cos. x = 1 - \frac{x^2}{1.2} + \frac{x^4}{1.2.3.4} - \frac{x^6}{1.2.3.4.5.6} + \&c. \quad (501)$$

93. Problem. To differentiate tang. x.

Solution. We have, by trigonometry and example 6, of art. 81,

$$d_c. \text{tang.} \, x = d_c. \frac{\sin. x}{\cos. x} = \frac{\cos. x \, d_c \sin. x - \sin. x \, d_c \cos. x}{\cos.^2 x},$$

whence, by (494), (496), and trigonometry,

$$d_{c_i}^1 . \, \text{tang.} \, x = \frac{\cos.^2 x + \sin.^2 x}{\cos.^2 x} = \frac{1}{\cos.^2 x}. \tag{502}$$

94. *Corollary.* Equations (502) and (496) give

$$d_c^2 . \, \text{tang.} \, x = - \frac{2 \, d_c \cos. \, x}{\cos.^3 x} = \frac{2 \sin. \, x}{\cos.^3 x} = \frac{2 \, \text{tang.} \, x}{\cos.^2 x}$$

or $d_c^2 . \, \text{tang.} \, x = 2 \, \text{tang.} \, x. \, d_c \, \text{tang.} \, x$

$d_c^3 \text{tang.} \, x = 2 \, \text{tang.} \, x \, d_c^2 \text{tang.} \, x + 2 \, (d_c \, \text{tang.} \, x)^2$

$d_c^4 \, \text{tang.} \, x = 2 \, \text{tang.} \, x \, d_c^3 \text{tang.} \, x + 6 \, d_c \, \text{tang.} \, x \, d_c^2 \text{tang.} \, x$

$d_c^5 \text{tang.} \, x = 2 \, \text{tang.} \, x \, d_c^4 \text{tang.} \, x + 8 \, d_c \, \text{tang.} \, x \, d_c^3 \text{tang.} \, x$

$$+ \, 6 \, (d_c^2 \, \text{tang.} \, x)^2$$

$$d_c^n \, \text{tang.} \, x = (d_c . \, \text{tang.} \, x + d_c . \, \text{tang.} \, x)^{n-1}. \tag{503}$$

in which the exponents are to be annexed to the $d_{c'}$; when the exponent is zero, as in the first and last terms, the $d_{c'}$ is omitted, and tang. x retained; and as the terms equally distant from the two extremities of the developed series are alike, they may be added together.

By making $x = 0$,

these equations give

$$d_c \, \text{tang.} \, x = 1$$

$$d_c^2 \, \text{tang.} \, x = 0$$

$$d_c^3 \, \text{tang.} \, x = 2$$

$$d_c^4 \, \text{tang.} \, x = 0$$

$$d_c^5 \, \text{tang.} \, x = 16.$$

Differential of negative sine and tangent.

Hence, by (447),

$$\text{tang. } x = x + \frac{\cdot 2}{1.2.3} x^3 + \frac{16}{1.2.3.4.5} x^5 + \&c.$$

$$= x + \tfrac{1}{3} x^3 + \tfrac{2}{15} x^5 + \&c. \qquad (504)$$

95. To differentiate arc sin. $x = \sin.^{[-1]} x$.

Solution. Let $y = \sin.^{[-1]} x$

or $\sin. y = x$

so that by differentiation

$$\cos. y. \, dy = dx = dy \sqrt{(1 - \sin.^2 y)} = dy \sqrt{(1 - x^2)}$$

and

$$d_c. \sin.^{[-1]} x = d_c. y = \frac{dy}{dx} = \frac{1}{\sqrt{(1-x^2)}} = (1-x^2)^{-\frac{1}{2}}. \quad (505)$$

96. *Corollary.* By the same process we should find

$$d_c. \cos.^{[-1]} x = d_c \text{ arc cos. } x = -(1-x^2)^{-\frac{1}{2}}. \qquad (506)$$

97. *Problem.* To differentiate $\text{tang.}^{[-1]} x = \text{arc tang.} x$.

Solution. Let $y = \text{tang.}^{[-1]} x$

or $\text{tang. } y = x,$

so that, by differentiation,

$$\frac{dy}{\cos.^2 y} = dx = dy \sec.^2 y = dy (1 + \text{tang.}^2 y) = dy (1 + x^2)$$

and

$$d_c \text{ tang.}^{[-1]} x = d_c y = \frac{dy}{dx} = (1 + x^2)^{-1}. \qquad (507)$$

98. *Scholium.* By applying Taylor's theorem, the values of $\sin.^{[-1]} x$, $\cos.^{[-1]} x$, and $\tan.^{[-1]} x$, might easily be found expressed in series ; but they may more easily be found by the following process.

99. *Problem.* To develop $\sin.^{[-1]} x$ in a series arranged according to powers of x.

Solution. Suppose the series to be

$$\sin.^{[-1]} x = C_0 + C_1 x + C_2 x^2 + C_3 x^3 + C_4 x^4 + \&\text{c}. \quad (508)$$

in which the number below the coefficient denotes the power of x, which it multiplies. We find, then, by differentiation, (505), and the binomial theorem,

$$(1 - x^2)^{-\frac{1}{2}} = C_1 + 2 C_2 x + 3 C_3 x^2 + 4 C_4 x^3 + \&\text{c}.$$

$$= 1 + \tfrac{1}{2} x^2 + \frac{1.3}{2.4} x^4 + \frac{1.3.5}{2.4.6} x^6 + \&\text{c}.$$

whence

$$C_1 = 1$$

$$C_3 = \tfrac{1}{3} \cdot \frac{1.}{2.} = \tfrac{1}{6}$$

$$C_5 = \tfrac{1}{5} \cdot \frac{1.3}{2.4} = \tfrac{3}{40}$$

and, $C_n = 0$

when n is even, and if in (508) we put

$$x = 0$$

it becomes

$$C_0 = 0.$$

Hence, by substitution,

$$\sin.^{[-1]} x = x + \tfrac{1}{6} x^3 + \tfrac{3}{40} x^5 + \&\text{c}. \quad (509)$$

100. *Corollary.* In the same way, if we put

$$\cos.^{[-1]} x = C_0 + C_1 x + C_2 x^2 + \&\text{c.} \qquad (510)$$

we find all the coefficients except C_0 to be the negative of those for $\sin.^{[-1]} x$, and if in (510) we put

$$x = 0$$

we find

$$C_0 = \text{arc. cos. } 0 = \tfrac{1}{4}\pi,$$

whence

$$\cos.^{[-1]} x = \tfrac{1}{4}\pi - x - \tfrac{1}{6} x^3 - \tfrac{3}{40} x^5 - \&\text{c.} \qquad (511)$$

101. *Corollary.* If in (511) we put

$$x = 1$$

we have

$$\cos.^{[-1]} 1 = 0 = \tfrac{1}{4}\pi - 1 - \tfrac{1}{6} - \tfrac{3}{40} - \&\text{c.} \qquad (512)$$

and if (512) is subtracted from (511), term from term, it gives

$$\cos.^{[-1]} x = (1-x) + \tfrac{1}{6}(1-x^3) + \tfrac{3}{40}(1-x^5) + \&\text{c.} \qquad (513)$$

102. *Problem.* To develop $\tan.^{[-1]} x$ in series, according to powers of x.

Solution. Suppose the series to be

$$\tan.^{[-1]} x = C_0 + C_1 x + C_2 x^2 + \&\text{c.} \qquad (514)$$

we find, by differentiation, (507), and the binomial theorem,

$$(1+x^2)^{-1} = C_1 + 2 C_2 x + 3 C_3 x^2 + 4 C_4 x^3 + \&\text{c.}$$
$$= 1 - x^2 + x^4 - x^6 + \&\text{c.}$$

whence

$$C_1 = 1$$
$$C_3 = -\tfrac{1}{3}$$
$$C_5 = \tfrac{1}{5}$$

and when n is odd

$$C_n = \pm \frac{1}{n},$$

18

Differentials of circular functions.
---

the upper sign being used when $n + 1$ is divisible by 4, and the lower sign when $n - 1$ is divisible by 4; but when n is even or zero,

$$C_n = 0.$$

Hence, by substitution,

$$\tan.^{[-1]} x = x - \tfrac{1}{3} x^3 + \tfrac{1}{5} x^5 - \tfrac{1}{7} x^7 + \&c. \qquad (515)$$

103. EXAMPLES.

1. Differentiate sec. x.　　　*Ans.*　tang. x . sec. $x \, dx$.

2. Differentiate sin.$^n x$.　　　*Ans.*　n sin.$^{n-1} x$ cos. $x \, dx$.

3. Differentiate log. cos. x.　　*Ans.*　— tang. $x \, dx$.

4. Differentiate log. $\sqrt{\left(\dfrac{1 + \sin. x}{1 - \sin. x}\right)}$.　　*Ans.*　sec. $x \, dx$.

5. Differentiate $\dfrac{1}{\sqrt{(a^2 - b^2)}}$ cos.$^{[-1]} \left(\dfrac{b + a \cos. x}{a + b \cos. x}\right)$.

$$\text{\textit{Ans.}} \quad \frac{d x}{a + b \cos. x}.$$

6. Differentiate cot. x.　　　*Ans.*　$- \dfrac{d x}{\sin.^2 x}$.

CHAPTER VII.

INDETERMINATE FORMS.

104. *Problem.* To find the value of a fraction when its numerator and denominator both vanish for a given value of the variable, of which they are both continuous functions.

Solution. Let the fraction be

$$\frac{F. x}{f. x} \tag{516}$$

and let x_0 be the value of x, for which the terms vanish; let

$$h = x - x_0, \text{ or } x = x_0 + h, \tag{517}$$

the given fraction becomes then

$$\frac{F. (x_0 + h)}{f. (x_0 + h)} \tag{518}$$

which is a fraction, both whose terms vanish for

$$h = 0 ; \tag{519}$$

so that in (456), h being the variable instead of x, we have, if all the differential coefficients of the terms vanish, up to the nth, when

$$h = 0,$$

$$\frac{F. (x_0 + h)}{f. (x_0 + h)} = \frac{d_c^n. F. (x_0 + \theta_n h)}{d_c^n. f. (x_0 + \theta_n h)}, \tag{520}$$

Fraction of which both terms vanish.

---

which when $h = 0$

becomes $\dfrac{F.\,x_0}{f.\,x_0} = \dfrac{d_c^n.\,F.\,x_0}{d_c^n.\,f.\,x_0}$;   (521)

so that the value of the given fraction is obtained by differentiating its numerator and denominator, until a differential coefficient is obtained, which does not vanish for the given value of the variable.

105. EXAMPLES.

1. Find the value of $\dfrac{e^x - e^{-x}}{\sin.\,x}$, when $x = 0$.

Solution. We have

$$\frac{e^0 - e^{-0}}{\sin.\,0} = \frac{d_c.\,e^0 - d_c.\,e^{-0}}{d_c.\,\sin.\,0} = \frac{e^0 + e^{-0}}{\cos.\,0} = \frac{1 + 1}{1} = 2.$$

2. Find the value of $\dfrac{\sin.\,(x^2)}{x}$, when $x = 0$.   *Ans.* 0.

3. Find the value of $\dfrac{\sin.\,x}{x^2}$, when $x = 0$.   *Ans.* ∞.

4. Find the value of $\dfrac{\log.\,x}{x - 1}$, when $x = 1$.   *Ans.* 1.

5. Find the value of $\dfrac{x^n - a_s^n}{x - a}$, when $x = a$.   *Ans.* $n\,a^{n-1}$.

6. Find the value of $\dfrac{1 - \cos.\,x}{x^2}$, when $x = 0$.   *Ans.* $\frac{1}{2}$.

7 Find the value of $\dfrac{x - \sin.\,x}{x^3}$, when $x = 0$.   *Ans.* $\frac{1}{6}$.

Fraction of which both terms are infinite.

8. Find the value of $\dfrac{e^x - e^{-x} - 2\,x}{x - \text{sin.}\,x}$, when $x = 0$.

$Ans.\ 2.$

9. Find the value of $\dfrac{e^x - e^{-x} - 2\,\text{sin.}\,x}{x^3}$, when $x = 0$.

$Ans.\ \frac{1}{3}.$

10. Find the value of $\dfrac{\text{sin.}\,m\,\varphi}{\text{sin.}\,\varphi}$, when $\varphi = 0$. $Ans.\ m.$

11. Find the value of $\dfrac{\text{sin.}\,m\,\varphi}{\text{sin.}\,\varphi}$, when $\varphi = \pi$, and m is an integer.

$Ans.$ $- m\,\text{cos.}\,m\,\pi$, so that when m is even, the answer is $- m$; and when m is odd, the answer is m.

106. *Problem.* To find the value of a fraction, when both its terms become infinite for a given value of the variable.

Solution. Let the numerator of the fraction be Y and the denominator y, the fraction is

$$\frac{Y}{y} = \frac{y^{-1}}{Y^{-1}}. \tag{522}$$

But when y and Y are infinite, their reciprocals, y^{-1} and Y^{-1}, vanish, and we have by the preceding art.

$$\frac{Y}{y} = \frac{d_e.\,y^{-1}}{d_e.\,Y^{-1}} = \frac{-y^{-2}d_e\,y}{-Y^{-2}d_e^i\,Y} = \frac{Y^2\,d_e\,y}{y^2\,d_e^i\,Y}$$

and, dividing by $\dfrac{Y^2}{y^2}$

$$\frac{y}{Y} = \frac{d_e\,y}{d_e\,Y}, \quad \text{or} \quad \frac{Y}{y} = \frac{d_e\,Y}{d_e\,y}, \tag{523}$$

18*

---

so that the value of the fraction may be found by differentiating both its terms; and if both the terms of the fraction thus obtained are infinite or zero, the differentiation may be continued, until a fraction is obtained, of which both terms are not infinite or zero ; and equation (521) applies to the present case as well as to the preceding one.

107. Examples.

1. Find the value of $\dfrac{\log. x}{\cot. x}$, when $x = 0$.

Solution. We have

$$\frac{\log. 0}{\cot. 0} = \frac{d_c. \log. 0}{d_c. \cot. 0} = -\frac{\sin.^2 0}{0} = -2 \sin. 0 \cos. 0 = 0.$$

2. Find the value of $\dfrac{\cot. m \varphi}{\cot. \varphi}$, when $\varphi = 0$.   *Ans.* $\dfrac{1}{m}$.

108. *Problem.* To find the value of a product of two factors, when one of the factors become infinite and the other factor becomes zero for a given value of the variable.

Solution. Let y and Y be the two factors, and we have for the given product

$$y\, Y = \frac{Y}{y^{-1}} \tag{524}$$

so that it is equal to a fraction, of which both the terms

are infinite, or both are zero, and the value of this fraction may be found by the preceding articles.

109. EXAMPLES.

1. Find the value of $x^a e^{-x}$, when $x = \infty$.   *Ans.* 0.

2. Find the value of $x (\log. x)^n$, when $x = 0$.   *Ans.* 0.

3. Find the value of $x^a \log. x$, when $x = 0$.   *Ans.* 0.

4. Find the value of $x \cotan. x$, when $x = 0$.   *Ans.* 1.

110. *Problem.* To find the value of a power, when the exponent and the root are both such functions of a variable, that they assume, for a given value of the variable, one of the forms 0^0, ∞^0, or 1^∞.

Solution. Let the power be

$$z = Y^y \qquad (525)$$

and we have, by logarithms,

$$\log. z = y \log. Y, \qquad (526)$$

so that in either of the given cases $\log. z$ is equal to a product, of which one of the factors is zero, while the other is infinite; its value may therefore be found by the preceding articles; and when its value is found, we have

$$z = e^{\log. z} \qquad (527)$$

or $\qquad Y^y = e^{y \log. Y}. \qquad (528)$

Infinite or zero powers.

111. EXAMPLES.

1. Find the value of x^x, when $x = 0$.

Solution. Since $0 \log. 0 = 0$

we have $0^0 = e^0 = 1$.

2. Find the value of $(e^x - 1)^x$, when $x = 0$.   *Ans.* 1.

3. Find the value of $\cot.^\varphi \varphi$, when $\varphi = 0$.   *Ans.* 1.

4. Find the value of $\cos. \varphi^{\cot. \varphi}$, when $\varphi = 0$.   *Ans.* 1.

CHAPTER VIII.

MAXIMA AND MINIMA.

112. A value of a function, which is greater than those immediately preceding and following it, is called a *maximum* of the function; while a value, which is less than those adjacent to it, is called a *minimum*.

113. *Problem.* Find the maxima and minima of a continuous function of one variable.

Solution. Let f. be the given function, of which all the differential coefficients, inferior to the nth, vanish, for a given value x_0 of the variable. Then

$$f.(x_0 + h) - f.x_0 \qquad (529)$$

is a function of h, which vanishes with h; and its differential coefficients, inferior to the nth, also vanish with h, for they are, when h is zero,

$$d_c^m . f.(x_0 + h) = d_c^m . f.x_0 = 0. \qquad (530)$$

Moreover h^n is a function of h, all the differential coefficients of which, inferior to the nth, vanish with h, and the nth differential coefficient is

$$d_c^n . h^n = 1.2.3....n. \qquad (531)$$

If now, in (456), these two functions are substituted, and if h is regarded as the variable, (456) becomes

Rule for maxima and minima.

$$\frac{f.(x^0 + h) - f.x^0}{h^n} = \frac{d_c^n.f.(x_0 + \theta_n h)}{1.2.3.\ldots n} \quad (532)$$

But if h is taken infinitely small, $x_0 + \theta_n h$ differs infinitely little from x_0, and the $\theta_n h$ may be neglected, so that (532), multiplied by h^n, becomes, by transposition,

$$f.(x_0 + h) = f.x_0 + \frac{h^n}{1.2.3.\ldots n} d_c^n.f.x_0. \quad (533)$$

By changing h into $- h$, (533) becomes

$$f.(x_0 - h) = f.x_0 + \frac{(-h)^n}{1.2.3.\ldots n} d_c^n.f.x_0. \quad (534)$$

If, then, n is odd, of the values of the given function, which are adjacent to $f.x_0$, one is greater, while the other is less than $f.x_0$, so that this value does not correspond to a maximum or minimum.

But if n is even, the values which are adjacent to $f.x_0$ are both greater than $f.x_0$, when the nth differential coefficient is positive; and they are both less than $f.x_0$, when this coefficient is negative.

Hence, to obtain the maxima and minima of a given function, find the values of the variable which reduce the first differential coefficient to zero. Each of these values of the variable must be substituted in the succeeding differential coefficients, until one is arrived at, which does not vanish.

If the first differential coefficient which does not vanish is even and positive, the corresponding value of the function is a minimum; but if the coefficient is even and negative, the value of the function is a maximum.

Case when differential coefficient is infinite.

114. *Scholium.* When the function, varying with the increase of its variable, passes through a maximum or minimum, its difference must change its sign. If this difference is continuous, it can only change its sign by passing through zero ; but if it is discontinuous, it may also change its sign by passing through infinity, and the first differential coefficient which does not vanish must in this case be infinite.

The case, therefore, in which the first differential coefficient, which does not vanish, is infinite, deserves particular examination ; and all other cases of discontinuity are to be considered by themselves, but they are rarely of much interest.

115. EXAMPLES.

1. Find the maxima and minima of the function $x^3 + ax + b$.

Solution. The differential coefficients are

$$d_c . (x^3 + ax + b) = 3x^2 + a$$

$$d_c^2 . (x^3 + ax + b) = 6x$$

$$d_c^3 . (x^3 + ax + b) = 6.$$

The first coefficient is zero when

$$x = \pm \sqrt{(-\tfrac{1}{3} a)},$$

so that there is neither maximum nor minimum, unless a is zero or negative. For this value of x, the second coefficient becomes

$$\pm 6 \sqrt{-\tfrac{1}{3} a},$$

which, when a is negative, is positive for the positive value of

x, and negative for the negative value of x. The corresponding values of the function are

$$b \pm \tfrac{2}{3} a \sqrt{(-\tfrac{1}{3} a)}.$$

But when a is zero, the second differential coefficient also
vanishes, and the third does not, so that there is neither maximum nor minimum.

2. Find the maxima and minima of the function

$$a x^2 + b x + c.$$

Ans. The value $\dfrac{4 a c - b^2}{4 a}$ is a maximum when a is

negative, and a minimum when a is positive; the corresponding value of x is

$$-\frac{b}{2 a}.$$

3. Find a maximum or minimum of the function

$$e^x + e^{-x} + 2 \cos. x.$$

Ans. The value 4 is a minimum, which corresponds to
the value

$$x = 0.$$

4. Find a maximum or minimum of the function

$$e^x + e^{-x} - x^2.$$

Ans. The value 2 is a minimum, which corresponds to the
value

$$x = 0.$$

5. Divide a number into two such parts that their product
may be a maximum or minimum.

Ans. The product is a maximum, when the two parts are
equal.

6. Divide a number into two such parts that the product of the mth power of the one by the nth power of the other may be a maximum or a minimum, m and n being positive.

Ans. The product is a maximum, when the parts are in the same ratio as the exponents of their powers.

7. Inscribe in the triangle ABC (fig. 8.), the greatest or the least possible rectangle $DEFG$.

Solution. Using the notation of art. 41, example 2, we have

$$\text{the surface } DEFG = \frac{b\,x\,(h-x)}{h},$$

which is a maximum, when

$$x = \tfrac{1}{2}\,h.$$

8. Through a given point C (fig. 15.) draw a line BCD, so that the surface of the triangle ABD, intercepted between the lines AB and AD, may be a maximum or a minimum.

Solution. This surface is, by art. 42, example 7,

$$\tfrac{1}{2}\,x\,y\,\sin.\,A = \tfrac{1}{2}\,(a\,y + b\,x)\,\sin.\,A.$$

the first and second differential coefficients of which are

$$0 = \tfrac{1}{2}\,(y + x\,d_c y)\,\sin.\,A = \tfrac{1}{2}\,(a\,d_c y + b)\,\sin.\,A.$$

$$(d_c y + \tfrac{1}{2}\,x\,d_c^2 y)\,\sin.\,A = \tfrac{1}{2}\,a\,d_c^2 y\,\sin.\,A.$$

Hence

$$d_c y = \frac{b - y}{x - a}$$

$$0 = b\,x - a\,y,$$

$$y = 2\,b, \quad x = 2\,a, \quad d_c^2 y = \frac{2\,b}{a^2},$$

and the surface is a minimum.

19

9. Find on the circle (fig. 33.) referred to the centre B, as the origin of rectangular coördinates, the point M, for which the product of the coördinates is a maximum or a minimum.

Ans. The product is a maximum when $y = x$.

10. Find on the ellipse (fig. 35.), referred to its centre as the origin, its axes as the axes of coördinates, the point M, for which the product of the coördinates is a maximum or a minimum.

Ans. The product is a maximum, when the coördinates are in the ratio of the axes.

116. When the function, of which the maxima and minima are to be found, is a product or quotient, the solution is often simplified by finding the maxima and minima of its logarithm, which evidently correspond to those of the function.

EXAMPLES.

1. Find the maxima and minima of the function $x^{-a} e^{bx}$, when a is positive.

Solution. We have

$$\log. (x^{-a} e^{bx}) = - a \log. x + bx$$

$$d_e . \log. (x^{-a} e^{bx}) = - \frac{a}{x} + b$$

$$d_e^2 . \log. (x^{-a} e^{bx}) = \frac{a}{x^2}$$

so that the value

$$x = \frac{a}{b}$$

corresponds to a maximum.   The corresponding value of the given function is

$$\left(\frac{b\,e}{a}\right)^{a}.$$

2. Find the maxima and minima of the function

$$(x - a)\,(x - b)\,x^{-2}.$$

Ans.   There is a minimum, when

$$x = \frac{2\,a\,b}{a + b}.$$

CHAPTER IX.

CONTACT.

117. When two curves meet or cut at a given point, and, at any infinitesimal distance of the first order from this point, are at a distance apart, which is an infinitesimal of the $n + 1$st order, they are said to be in *contact* at this point, and their contact is *of the nth order*.

When one of the curves is of the first degree or the straight line, it is the ordinary *tangent*.

When n is zero, there is no contact, but only an intersection.

118. *Theorem.* When two curves are in contact, the portion which is intersected between them upon a line, drawn at an infinitesimal distance from the point of contact, and inclined by a finite angle to the directions of the curves, is of the same order of infinitesimals with the distances of the curves apart.

Proof. Let $M'M_0M$ and $M_1'M_0M_1$ (fig. 49.) be the curves, A the point of contact, MN their distance apart, and MM_1 the intercepted portion of the line PM. The line M_1N may be regarded as a straight line, and the angle MNM_1 as a right angle; so that, we have

$$M M_1 = M N \times \sec. N M M_1.$$

But sec. $N M M_1$ is finite, as long as the angle $M M_1 N$, by which $M N M_1$ differs from a right angle, is finite; and, therefore, by art. 31, $M M_1$ is of the same order with $M N$.

119. *Problem. Find the algebraic conditions which denote that two given curves have a contact of the nth order.*

Solution. Let the equations of the two given curves $M' M_0 M$, $M'_1 M_0 M_1$ (fig. 49.), referred to rectangular coördinates, be, respectively,

$$y = f. x \tag{535}$$

$$y = f_1 . x. \tag{536}$$

Then if A is the origin and M_0 the point of contact, we have at this point

$$A P_0 = x_0, \quad P_0 M_0 = y_0$$

or $\quad f. x_0 = f_1 . x_0$ or $f. x_0 - f_1 . x_0 = 0,$ \tag{537}

and if $\quad\quad\quad P P_0 = h$

we have $\quad M M_1 = f. (x_0 + h) - f_1 . (x_0 + h).$ \tag{538}

If, now, $f. x_0 - f_1 . x_0$ is such a function of h that all its differential coefficients inferior to the mth are zero, we have, by (533) and (537), substituting $f. - f_1.$ for $f.$

$$M M_1 = \frac{h^m}{1.2.3.\ldots m} d_c^m. (f. x_0 - f_1 . x_0) \tag{539}$$

and if we put

$$X_0 = \frac{d_c^m. (f. x_0 - f_1 . x_0)}{1.2.3.\ldots m} = \frac{d_c^m. f. x_0 - d_c^m f_1 . x_0}{1.2.3.\ldots m} \tag{540}$$

19*

we have

$$MM_1 = X_0 \cdot h^m, \tag{541}$$

so that if h is of the first order, MM_1 is of the mth order. But MM_1 is to be of the $(n+1)^{st}$ order, and therefore we must have

$$m = n+1, \ MM_1 = X_0 \cdot h^{n+1} \tag{542}$$

or

$$X_0 = \frac{d_c^{n+1} f \cdot x_0 - d_c^{n+1} \cdot f_1 \cdot x_0}{1.2.3 \ldots \ldots (n+1)} \tag{543}$$

and all the differential coefficients of $f \cdot x_0 - f_1 \cdot x$, inferior to the $(n+1)^{st}$, must be zero; that is, we must have, when m is less than $n+1$,

$$d_c^m f \cdot x_0 - d_c^m f_1 \cdot x_0 = 0 \tag{544}$$

or

$$d_c^m f \cdot x_0 = d_c^m f_1 \cdot x_0. \tag{545}$$

120. In the same way, it may be proved that if the equations of the two curves are expressed in the polar coördinates of art. 45; so that they are

$$r = F \cdot \varphi, \ r = F_1 \cdot \varphi \tag{546}$$

we must have, when m is less than $n+1$, for the point r_0, φ_0 of contact,

$$d_c^m F \cdot \varphi_0 = d_c^m F_1 \cdot \varphi_0. \tag{547}$$

121. *Theorem.* Two curves, which are in contact, cross each other at the point of contact, when the contact is of an even order; but, if the order of the contact is of an uneven order, they do not cross.

Proof. For, when h is negative in (541), the sign of MM_1 is the same as when h is positive if n is uneven, but is the reverse if n is even ; that is, if n is even, the point M' (fig. 49.) is nearer AP than M_1', while the point M of the same curve with M' is farther from AP than M_1 of the same curve with M_1' ; and if n is uneven, the reverse is the case, as in (fig. 50.)

122. *Problem. Through a given point upon a given curve, to draw a tangent to the curve.*

Solution. Let x_0 and y_0 be the coördinates of the given point, and (535) the equation of the given curve. If τ is the angle which the tangent makes with the axis of x, its equation, since it passes through the given point, is by (162),

$$y - y_0 = \text{tang. } \tau \,(x - x_0).\qquad(548)$$

Hence, by differentiation,

$$d_{c.}\, y = \text{tang. } \tau = d_{c.}\, y_0 = d_{c.} f.\, x_0 \qquad(549)$$

and the equation of the tangent is

$$y - y_0 = d_{c.} f.\, x_0 \,(x - x_0).\qquad(550)$$

123. The projection PT' (fig. 51) of the tangent MT upon the axis of x is called the *subtangent*, and if we put

$$AT = x'$$

the coördinates of T are

$$y = 0 \text{ and } x = x',$$

so that

$$PT = x_0 - x' = \frac{y_0}{d_{c.} f.\, x_0} = \frac{y_0}{d_{c.}\, y_0} = \frac{f.\, x_0}{d_{c.} f.\, x_0}.\qquad(551)$$

124. *Corollary.* When the equation of the curve is expressed in polar coördinates, as in (546), the tangent can be found
by means of (117). Thus we have

$$\tau = \tfrac{1}{2} \pi - \alpha, \text{ or } \alpha = \tfrac{1}{2} \pi - \tau \qquad (552)$$

$$\varphi + \alpha = \tfrac{1}{2} \pi - \tau + \varphi = \tfrac{1}{2} \pi - (\tau - \varphi)$$

$$r \cos. (\varphi + \alpha) = r \sin. (\tau - \varphi) = p, \qquad (553)$$

and by logarithms

$$\log. r + \log. \sin. (\tau - \varphi) = \log. p. \qquad (554)$$

The differential of which is, by transposition,

$$\frac{d_c r}{r} = \frac{\cos. (\tau - \varphi)}{\sin. (\tau - \varphi)} = \frac{1}{\tang. (\tau - \varphi)} = \cotan. (\tau - \varphi). \ (555)$$

But in (fig. 51.), we have

$$\varphi = MAP, \ \tau = MTP, \qquad (556)$$

and if we put

$$\varepsilon = AMT = \tau - \varphi \qquad (557)$$

ε is the angle which the curve makes with the radius vector
at the point M, so that

$$\cotan. \varepsilon = \frac{d_c r}{r} = d_c \log. r = d_c \log. F. \varphi_0 \qquad (558)$$

and the equation of the tangent at the point $r_0 \ \varphi_0$ is

$$r \sin. (\tau - \varphi) = r_0 \sin. \varepsilon. \qquad (559)$$

125. The perpendicular MI to the tangent at the
point of contact is called *the normal* to the curve.

Normal.

If ν is the angle which the normal makes with the axis of x, we have then, by (549),

$$\nu = \tfrac{1}{2}\pi + \tau \qquad (560)$$

$$\text{tang. } \nu = - \text{cotan. } \tau = - \frac{1}{\text{tang.}\,\tau} = - \frac{1}{d_c y_0}, \qquad (561)$$

and the equation of the normal is

$$y - y_0 = \text{tang. } \nu \,(x - x_0) = - \frac{1}{d_c y_0}(x - x_0). \qquad (562)$$

126. The projection PI of the normal MI upon the axis of x is called the *subnormal*, so that its value, found like that of the subtangent, is

$$AI - x_0 = y_0\, d_c y_0 = f.\,x_0 \,.\, d_c f.\,x_0. \qquad (563)$$

127. *Corollary.* The lengths of the tangent and normal, found from the right triangles MTP and MPI, are

$$T = MT = \sqrt{(MP^2 + PT^2)} = \frac{y_0}{d_c y_0}\sqrt{[1 + (d_c y_0)^2]} \quad (564)$$

$$N = MI = \sqrt{(MP^2 + PI_2)} = y_0\sqrt{[1 + (d_c y_0)^2]}$$
$$= T\, d_c y_0. \qquad (565)$$

128. The tangent is called an asymptote, when the point of contact is at an infinite distance from the origin.

129. *Scholium.* It must not be overlooked that in finding $d_c y$, x has been regarded as the independent variable. But if it were not so, and if some other variable, as u, were the independent variable; then, denoting by $d_{c.x}.,\ d_{c.u}.,$ the differential

coefficients taken on the supposition that x, u are respectively the independent variables, we have, at once,

$$d_{c.x} y = \frac{d y}{d x} = \frac{d_{c.u} y}{d_{c.u} x}. \tag{566}$$

130. If all the terms of the equation of the given curve were transposed to its first member, and if V were this first member, the equation would be

$$V = 0, \tag{567}$$

whence by differentiation, putting V_0 for the value of V at the point of contact

$$d_{c.x} V_0 \, d x_0 + d_{c.y} V_0 \, d y = 0, \tag{568}$$

from which the value of $\frac{d y}{d x}$, being found and substituted in (548), gives by reduction for the equation of the tangent

$$d_{c x.0} V_0 . (x - x_0) + d_{c y.0} V_0 (y - y_0) = 0, \tag{569}$$

so that the equation of the tangent may be found by differentiating the equation of the curve and substituting $x - x_0$ and $y - y_0$ for $d x$ and $d y$ respectively.

131. Examples.

1. Find the tangent, normal, &c. of the circle.

Solution. The differential of equation (58) of the circle gives for any point x_0, y_0,

$$x_0 + y_0 \, d_c y_0 = 0,$$

$$\text{tang. } \tau = - \frac{x_0}{y_0}, \quad \text{tang. } \nu = \frac{y_0}{x_0}.$$

The equation of the tangent, reduced by (58), is

$$y_0 y + x_0 x = y_0^2 + x_0^2 = R^2,$$

and that of the normal is

$$x_0 y - y_0 x = 0,$$

so that the normal, by art. 109 of Book I., passes through the centre, and is the radius. We have also

$$\text{the subtangent} = -\frac{y_0^2}{x_0} = -\frac{R^2 - x_0^2}{x_0}$$

$$\text{the subnormal} = -x_0$$

$$\text{the tangent} = \frac{R\,y_0}{x_0}, \text{ the normal} = R.$$

Again, from equation (56) we find

$$\cot. \; \varepsilon = d_c \log. \; R = 0$$

$$\varepsilon = 90°;$$

that is, the radius vector drawn from the centre is perpendicular to the tangent.

2. Find the tangent, normal, &c. of the ellipse, of which A and B are the semiaxes.

$$Ans. \quad \text{tang.}\; \tau = -\frac{B^2 x_0}{A^2 y_0}, \; \text{tang.}\; \nu = \frac{A^2 y_0}{B^2 x_0}.$$

The equation of the tangent is

$$A^2 y_0 y + B^2 x_0 x = A^2 B^2;$$

that of the normal is

$$-B^2 x_0 y + A^2 y_0 x = (A^2 - B^2) x_0 y_0.$$

$$\text{The subtangent} = -\frac{A^2 y_0^2}{B^2 x_0} = -\frac{A^2 - x_0^2}{x_0} = x_0 - \frac{A^2}{x_0}.$$

The subnormal $= -\dfrac{B^2\, r_0}{A^2}$.

When the focus F (fig. 34) is the origin

$$\text{cotan.}\ \varepsilon = -\frac{c\sin.\ \varphi_0}{A - c\cos.\ \varphi_0} = \frac{-c\, r\sin.\ \varphi_0}{B^2} = -\frac{c\, y_0}{B^2}.$$

When the focus F' is the origin

$$\text{cotan.}\ \varepsilon' = \frac{c\, y_0}{B^2} = -\text{cotan.}\ \varepsilon$$

$$\varepsilon' = -\ \varepsilon.$$

Corollary. The lines FM, $F'M$, (fig. 52.) drawn from the foci to any point M of the ellipse make equal angles FMt, F_1Mt, with the tangent at the point M. Hence a tangent may be drawn to the ellipse by bisecting the angle FMF_1 by the line TMt, which will be the tangent required.

3. The tangent, normal, &c. of the hyperbola are found from those of the ellipse by changing the sign of B^2. *by jo. 66.*

We hence find for PT (fig. 51.) in this case

$$PT = x_0 - \frac{A^2}{x_0}$$

$$AT = AP - PT = \frac{A^2}{x_0},$$

so that for the asymptote we have

$$x = \infty, \quad AT = 0,$$

$$\text{cotan.}\ \varepsilon = -\infty, \quad \varepsilon = \tau - \varphi = 0, \quad \tau = \varphi.$$

But, in this case, we have by cor. 1, of the hyperbola, B. I. § 98,

$$r = \infty, \quad \cos.\ \varphi = \cos.\ \varphi_0 = \frac{A}{C}$$

and the lines EAE_1', E_1AE' (fig. 36.) are asymptotes of the hyperbola.

4. Find the tangent, normal, &c. of the parabola, whose parameter is $4\,p$.

Ans. tang. $\tau = \dfrac{2\,p}{y_0}$, tang. $\nu = -\dfrac{y_0}{2\,p}$.

The equation of the tangent is

$$y_0\, y = 2\,p\,(x + x_0),$$

that of the normal is

$$2\,p\,y + y_0\,x = (2\,p + x_0)\,y_0.$$

The subtangent $= 2\,x_0$,

the subnormal $= 2\,p$,

and when the focus is the origin

$$\text{cotan. } \epsilon = -\frac{\sin. \varphi}{1 - \cos. \varphi} = -\frac{2\sin. \frac{1}{2}\varphi \cos. \frac{1}{2}\varphi}{2\cos. \frac{1}{2}\varphi} = -\text{cotan.} \tfrac{1}{2}\varphi$$

$$\epsilon = -\tfrac{1}{2}\varphi = \tau - \varphi$$

$$\tau = \tfrac{1}{2}\varphi\,;$$

that is, if (fig. 39.) MT is the tangent, we have

$$MFP = FMQ = \varphi = 2\,FMT,$$

so that a tangent may be drawn to the parabola by bisecting the angle FMQ.

5. Find the tangent, normal, &c. of the cycloid.

Solution. Taking θ as the independent variable, we have by (549) and (566)

20

$$\text{tang. } \tau = d_{ex} y = \frac{d_c \cdot y}{d_c \cdot x} = \frac{\sin \theta}{1 - \cos \theta} = \text{cotan. } \tfrac{1}{2} \theta$$

$$\tau = \tfrac{1}{2} \pi - \tfrac{1}{2} \theta, \; \nu = \pi - \tfrac{1}{2} \theta.$$

But if (fig. 41.) MB is joined, the angle MBX is measured by a semicircumference diminished by half the arc MB, that is,

$$MBX = \pi - \tfrac{1}{2} MCB = \pi - \tfrac{1}{2} \theta = \nu,$$

so that MB is the normal to the cycloid, and MT, which is drawn perpendicular to MB, is the tangent.

The subtangent $= 2 R \sin.^3 \tfrac{1}{2} \theta \sec. \tfrac{1}{2} \theta$

the subnormal $= R \sin. \theta = R \theta - x$

the tangent $= 2 R \sin.^2 \tfrac{1}{2} \theta \sec. \tfrac{1}{2} \theta$

the normal $= 2 R \cos. \tfrac{1}{2} \theta.$

6. Find the tangent of the spirals of equation (133).

Ans. For the tangent

$$\text{cotan. } \varepsilon = \frac{n}{\varphi_0}.$$

7. Find the tangent of the logarithmic spiral.

Ans. If the logarithms of equation (136) are taken in the system of which the base is a, (136) converted into Neperian logarithms is

$$\log. r = \log. a \cdot \varphi$$

and we have, for the cotangent,

$$\text{cotan. } \varepsilon = \log. a.$$

Order of contact of straight line.

8. Find the tangent and asymptote of the hyperbolic spiral.

Ans.   For the tangent

$$\text{cotan.}\, \varepsilon = -\frac{1}{\varphi_0}$$

$$\text{dist. of tang. from origin} = \frac{2\,\pi\,R}{\sqrt{(1+\varphi_0^2)}}$$

which for the asymptote become

$$\varphi_0 = 0,\ \varepsilon = \tau - \varphi_0 = 0,\ \tau = \varphi_0 = 0,$$

$$\text{dist. of asymp. from origin} = 2\,\pi\,R,$$

so that the asymptote is parallel to the polar axis.

132. Corollary. The straight line is completely determined by the condition, that the first differential coefficient of its ordinate is equal to that of the curve at the point of intersection ; and, therefore, the tangent has usually only a contact of the first order with a given curve ; so that, by art. 121, the tangent does not usually cross the curve at the point of contact.

133. Corollary. A point of one curve may be placed upon a point of another, and the two curves turned around upon this common pivot until their tangents coincide ; in this position, the two curves have evidently a contact of the first order. If now one of the curves is everywhere of the same curvature, that is, if it is a circle, the contact will remain of the same order, whichever of its points is brought to the point of contact ; but if it is any other curve, a point of it can usually be found, which, brought to the point of contact, will elevate the order of contact to the next higher order.

134. *Corollary.* By changing the dimensions of one of the curves, the order of contact can usually be increased by unity, for each of the constants which enter into the equation of this curve, and upon which its dimensions depend ; for each of these constants, regarded as an unknown quantity, may be determined so as to satisfy a new equation of those (545) or (547) upon which the order of contact depends.

Thus, a circle can usually be found, which has a contact of the second order, with a given curve at a given point ; a parabola, a cycloid, or a spiral of the form (133), (135), or (136), which has a contact of the third order; an ellipse or hyperbola, which has a contact of the fourth order, &c.

135. The differential of the arc of a curve is called its *element.*

Thus, if s denotes the arc, ds is the element of the curve. Hence in (fig. 53.) if

$$d x = P P' = MN,$$

we have then, by regarding MM' as a straight line,

$$d y = M'N, \ d s = MM' = \sqrt{(dx^2 + dy^2,)} \quad (570)$$

so that if x is the independent variable

$$d_e s = \sqrt{[1 + (d_e y)^2]} \quad (571)$$

and if s is the independent variable, using the notation of art. 129,

$$1 = \sqrt{[(d_{c.s}x)^2 + (d_{c.s}y)^2]}. \qquad (572)$$

136. *Corollary.* In the same way, if the radii vectores AM, AM' are drawn, and the arc MR described with R as a radius, we have

$$MAM' = d\varphi,\ M'R = AM' - AM = dr,\ MR = r\,d\varphi, \qquad (573)$$

and if MRM' is regarded as a right triangle

$$d s = \sqrt{(d r^2 + r^2 d \varphi^2)}, \qquad (574)$$

so that if φ is regarded as the independent variable

$$d_c s = \sqrt{[(d_c r)^2 + r^2]}. \qquad (575)$$

137. *Corollary.* The triangle $M'MN$ gives, by (560), x being regarded as the independent variable, when none is expressed in the formula

$$M'MN = \tau$$

$$\cos. \tau = \frac{dx}{ds} = \frac{1}{d_c' s} = d_{c.s}\, x = \sin. \nu \qquad (576)$$

$$\sin. \tau = \frac{dy}{ds} = \frac{d_c\, y}{d_c\, s} = d_{c.s}\, y = -\cos. \nu \qquad (577)$$

and by (564) and (565)

$$\text{tang.} = \frac{y\, d_{c'}\, s}{d_{c'}\, y} = y\, d_{c'\, y}\, s = \frac{y}{d_{c'.s}\, y} = \frac{y}{\sin. \tau} = -\frac{y}{\cos. \nu} \qquad (578)$$

$$\text{normal} = y\, d_c\, s = \frac{y}{d_{c'.s}\, x} = \frac{y}{\cos. \tau} = \frac{y}{\sin. \nu}. \qquad (579)$$

20*

138. *Corollary.* The triangle MRM' gives

$$MM'R = \varepsilon \tag{580}$$

$$\cos \varepsilon = \frac{dr}{ds} = \frac{d_c \cdot r}{d_c \cdot s} = d_{c'}, r \tag{581}$$

$$\sin \varepsilon = \frac{r \, d\varphi}{d s} = \frac{r}{d_c s} = r \, d_{c'}, \varphi, \tag{582}$$

φ being regarded as the independent variable, when none is expressed in the formula.

139. Two surfaces which intersect at a given point are said to be in contact, when their sections, found by any plane passing through this point, are in contact.

140. *Problem.* To find the algebraic conditions that two surfaces are in contact at a given point.

Solution. Let the surfaces be referred to rectangular coordinates, and let x_0, y_0, z_0 be the coördinates of the point of contact. Then the sections of the two surfaces, made by any plane, are found by equations [321 – 323], and since these sections are in contact, the values of $d_c y_1$, found on the hypothesis that x_1 is the independent variable, must be equal for the two surfaces at the point of contact. Hence the values of $d_c x_0$, $d_c y_0$, $d_c z_0$, found by differentiating the equations (321 – 323] must also be equal for the two surfaces, as well as

$$\frac{d_c y_0}{d_c x_0} = \frac{d y_0}{d x_0} = d_{c'} x \, y_0 \tag{583}$$

and
$$\frac{d_c z_0}{d_c x_0} = \frac{d z_0}{d x_0} = d_{c'} x \, z_0. \tag{584}$$

141. *Corollary.* If one of the surfaces is the tangent plane, its equation, since it passes through the point of contact, is by (197)

$$\cos. \alpha\,(x-x_0)+\cos.\beta\,(y-y_0)+\cos.\gamma\,(z-z_0)=0,\quad (585)$$

from which we find by differentiation

$$d_{c \cdot x}\, y = -\frac{\cos.\alpha}{\cos.\beta} = d_{c \cdot x}\, y_0 \qquad (586)$$

$$d_{c \cdot x}\, z = -\frac{\cos.\alpha}{\cos.\gamma} = d_{c \cdot x}\, z_0, \qquad (587)$$

which substituted in (585), divided by $\cos.\alpha$, give, for the equation of the tangent plane,

$$(x-x_0) - \frac{y-y_0}{d_{c \cdot x}\, y_0} - \frac{z-z_0}{d_{c \cdot x}\, z_0} = 0. \qquad (588)$$

142. *Corollary.* If all the terms of the equation of the given surface are transposed to the first member, and this first member, which is a function of x, y, z, represented by V, and if V becomes V_0 at the point of contact, the equation of the surface is

$$V = 0. \qquad (589)$$

The differential of this equation being taken on the hypothesis that z is constant, in order to find $d_{c \cdot x}\, y_0$, gives

$$d_{c \cdot x}\, V_0 . d\,x_0 + d_{c \cdot y}\, V_0 . d\,y_0 = 0, \qquad (590)$$

whence, by (586),

$$d_{c \cdot x}\, y_0 = \frac{d\,y_0}{d\,x_0} = -\frac{d_{c \cdot x}\, V_0}{d_{c \cdot y}\, V_0} = -\frac{\cos.\alpha}{\cos.\beta} \qquad (591)$$

or $\quad -\dfrac{1}{d_{c \cdot x}\, y_0} = \dfrac{\cos.\beta}{\cos.\alpha} = \dfrac{d_{c \cdot y}\, V_0}{d_{c \cdot x}\, V_0}. \qquad (592)$

Tangent plane.

In the same way we find

$$-\frac{1}{d_{c \cdot z} z_0} = \frac{\cos. \gamma}{\cos. \alpha} = \frac{d_{c \cdot z} V_0}{d_{c \cdot z} V_0},\qquad (593)$$

and these values, substituted in (588), give, by freeing from fractions,

$$d_{c \cdot z} V_0 (x - x_0) + d_{c \cdot y} V_0 (y - y_0) + d_{c \cdot z} V_0 (z - z_0) = 0. \quad (594)$$

This equation of the tangent plane compared with the complete differential of (589), which is

$$d_{c \cdot z} V_0 \, d x_0 + d_{c \cdot y} V_0 \, d y_0 + d_{c \cdot z} V_0 \, d z_0 = 0 \qquad (595)$$

shows that the equation of the tangent plane may be obtained from that of the surface by changing in the complete differential of the equation of the surface referred to the point of contact $d x_0$, $d y_0$, $d z_0$ respectively into $(x - x_0)$, $(y - y_0)$, $(z - z_0)$.

143. *Corollary.* The sum of the squares of (592) and (593) increased by unity is

$$\frac{\cos.^2 \alpha + \cos.^2 \beta + \cos.^2 \gamma}{\cos.^2 \alpha} = \frac{(d_{c \cdot z} V_0)^2 + (d_{c \cdot y} V_0)^2 + (d_{c \cdot z} V_0)^2}{d_{c \cdot z} V_0)^2} \quad (596)$$

or by (47) and putting

$$L_0 = \sqrt{[(d_{c \cdot z} V_0)^2 + (d_{c \cdot y} V_0)^2 + (d_{c \cdot z} V_0)^2]} \quad (597)$$

$$\cos. \alpha = \frac{d_{c \cdot z} V_0}{L_0}, \quad d_{c \cdot z} V_0 = L_0 \cos. \alpha \qquad (598)$$

$$\cos. \beta = \frac{d_{c \cdot y} V_0}{L_0}, \quad d_{c \cdot y} V_0 = L_0 \cos. \beta \qquad (599)$$

$$\cos. \gamma = \frac{d_{c \cdot z} V_0}{L_0}, \quad d_{c \cdot z} V_0 = L_0 \cos. \gamma. \qquad (600)$$

Normal to surface.

144. The perpendicular to the tangent plane drawn through the point of contact, is called the *normal* to the surface.

145. *Corollary.* The angles α, β, γ are, by (128), the angles which the normal makes with the axes; hence, by (124) or (598 – 600), the equations of the normal are

$$\frac{x - x_0}{\cos. \alpha} = \frac{y - y_0}{\cos. \beta} = \frac{z - z_0}{\cos. \gamma} \qquad (601)$$

or $\qquad \dfrac{x - x_0}{d_{c^.x} V_0} = \dfrac{y - y_0}{d_{c^.y} V_0} = \dfrac{z - z_0}{d_{c^.z} V_0}.$ $\qquad (602)$

146. EXAMPLES.

1. Find the equations of the tangent plane and of the normal to the sphere.

Solution. If the sphere is that of equation (62), we have

$$V = x^2 + y^2 + z^2 - R^2 = 0$$

$$d_{c^.x} V_0 = 2 x_0$$

$$d_{c^.y} V_0 = 2 y_0$$

$$d_{c^.z} V_0 = 2 z_0$$

$$L_0 = 2 \sqrt{(x_0^2 + y_0^2 + z_0^2)} = 2 R$$

$$\cos. \alpha = \frac{x_0}{R}, \ \cos. \beta = \frac{y_0}{R}, \ \cos. \gamma = \frac{z_0}{R}.$$

The equation of the tangent plane is, by reduction,

$$x_0 x + y_0 y + z_0 z = R^2$$

The equations of the normal are, by reduction,

$$\frac{x}{x_0} = \frac{y}{y_0} = \frac{z}{z_0},$$

so that by (128) it passes through the origin and is radius.

2. Find the equations of the tangent plane and of the normal to the ellipsoid of equation (335).

Ans. The equation of the tangent plane is, by reduction,

$$\frac{x_0 x}{A^2} + \frac{y_0 y}{B^2} + \frac{z_0 z}{C^2} - 1 = 0.$$

The equations of the normal are

$$A^2 \left(\frac{x}{x_0} - 1 \right) = B^2 \left(\frac{y}{y_0} - 1 \right) = C^2 \left(\frac{z}{z_0} - 1 \right).$$

3. Find the equations of the tangent plane to the cone of equation (364).

Ans. The equation of the tangent plane is

$$\frac{x_0 x}{A^2} + \frac{y_0 y}{B^2} - \frac{z_0 z}{C^2} = 0,$$

so that it passes through the origin.

4. Find the equations of the tangent plane and of the normal to the cylinder of equation (375).

Ans. The equation of the tangent plane is

$$B y_0 y + C z_0 z + M = 0,$$

so that it is perpendicular to the plane of $y z$. The equations of the normal are

$$x = x_0$$

$$C z_0 y - B y_0 z = (C - B) y_0 z_0.$$

Tangent and normal to paraboloid.

5. Find the equations of the tangent plane and of the normal to the paraboloid of equation (376).

Ans. The equation of the tangent plane is

$$2 B y_0^z y + 2 C z_0 z + H (x + x_0) = 0.$$

Those of the normal are

$$2 B y_0^z (x - x_0) = H (y - y_0)$$

$$C z_0 y - B y_0 z = (C - B) y_0 z_0.$$

6. Find the equations of the tangent plane and of the normal to the cylinder, of which the base is a parabola, and the equation is (384).

Ans. Put H cos. $\alpha + I$ sin. $\alpha = - 4 p\, C$, and omitting the numbers below the letters in (384); the equation of the tangent plane is

$$z_0 z = 2 p (x + x_0),$$

so that it is perpendicular to the plane of $x y$. The equations of the normal are

$$y = y_0$$

$$z_0 (x - x_0) = 2 p (z_c - z).$$

CHAPTER X.

CURVATURE.

147. The circle, which has the contact of the second order with a curve at a given point, coincides more nearly with the curve at that point than any other circle, and its curvature is therefore adopted as the *measure of the curvature* of the given curve at that point. It is hence called the *circle of curvature*, and its centre and radius are called, respectively, *the centre and radius of curvature*.

148. *Problem.* *To find the radius of curvature of a given curve at any point.*

Solution. Let ϱ be the required radius of curvature, ν the angle which the normal to the given curve makes with the axis of x, ν' the corresponding angle for the circle. Then, at the point x_0, y_0 of contact, we have

$$\nu_0 = \nu_0'. \tag{603}$$

But if s, s' are the arcs of the given curve, and of the circle, we have, by (576),

$$d s = \text{cosec.}\, \nu.\, d x, \quad d s' = \text{cosec.}\, \nu'.\, d x,$$

so that at the point of contact

$$d s_0 = d s_0'. \tag{604}$$

Radius of curvature.

But it is evident from (574) that, since the radius of the circle is constant, we have

$$d\,s_0 = d\,s_0' = \varrho\,d\,\nu_0'. \tag{605}$$

But the differential of (561) gives

$$d\nu_0 = \sin.^2 \nu_0\, d_{c \cdot z}^2 y_0\, dx_0, \quad d\nu_0 = \sin.'\nu_0'\, d_{c \cdot z}^2 y_0'\, dx_0, \tag{606}$$

and since the contact is of the second order

$$d_c^2 y_0 = d_c^2 y_0',$$

so that by (603) and (606)

$$d\nu_0 = d\nu_0', \tag{607}$$

which substituted in (605) gives

$$d\,s_0 = \varrho\,d\,\nu_0, \tag{608}$$

or omitting the cyphers below the letters,

$$\varrho = \frac{d\,s}{d\,\nu} = \frac{d_c\,s}{d_c\,\nu} = d_{c \cdot \nu}\,s. \tag{609}$$

149. *Corollary.* Equations (609), (571), (576), (577), (606), and (565) give

$$\varrho = \frac{d_{c \cdot z}\,s}{\sin.^2 \nu\, d_{c \cdot z}^2\,y} = \frac{(d_{c \cdot z}\,s)^3}{d_{c \cdot z}^2\,y} = \frac{\left[1 + (d_{c \cdot z}\,y)^2\right]^{\frac{3}{2}}}{d_{c \cdot z}^2\,y} \tag{610}$$

$$\varrho = \frac{(\sec. \tau)^3}{d_{c \cdot z}^2\,y} = \frac{N^3}{y^3\, d_{c \cdot z}^2\,y} = \frac{d\,s}{d\,\tau} = d_{c \cdot \tau}\,s. \tag{611}$$

150. *Corollary.* When the equation of the curve is given in polar coördinates, the radius of curvature may be found by means of equations (557) and (558). For these give

$$d\,\epsilon = - \sin.^2 \epsilon\, d.\left(\frac{d\,r}{r\,d\,\varphi}\right) \tag{612}$$

21

Radius of curvature.

$$d_\tau = d\varphi + d_\varepsilon = d\varphi - \sin.^2 {}_\varepsilon d. \left(\frac{dr}{r \, d\varphi}\right) \qquad (613)$$

$$\frac{1}{\varrho} = \frac{d\tau}{ds} = \frac{d\varphi}{ds} - \frac{\sin.^2 \varepsilon}{ds} d. \left(\frac{dr}{r d\varphi}\right). \qquad (614)$$

151. EXAMPLES.

1. Find the radius of curvature of the ellipse.

Solution. Equation (69) of the ellipse gives

$$d_{c \cdot x} y = -\frac{B^2 x}{A^2 y}$$

$$d_{c \cdot x}^2 y = -\frac{B^4}{A^2 y^3}$$

$$\varrho = \frac{(A^4 y^2 + B^4 x^2)^{\frac{3}{2}}}{A^4 B^4} = \frac{A^2 N^3}{B^4} = \frac{A^2 B^2}{(A^2 \sin.^2 \tau + B^2 \cos.^2 \tau)^{\frac{3}{2}}}.$$

This value of the radius of curvature is the same also for the hyperbola.

2. Find the radius of curvature of the parabola of equation of B. I. § 180.

$$\text{Ans.} \quad \frac{(y^2 + 4 p^2)^{\frac{3}{2}}}{4 p^2} = \frac{N^3}{4 p^2} = \frac{2 p}{\sin.^3 \tau}.$$

3. Find the radius of curvature of the cycloid.

$$\text{Ans. } 4 R \sin. \tfrac{1}{2} \theta = 4 R \cos. \tau = 2 N.$$

4. Find the radius of curvature of the spiral of Archimedes.

$$\text{Ans.} \quad \frac{R (1 + \varphi^2)^{\frac{3}{2}}}{2 (2 \pi + \varphi^2)}.$$

Evolute and involute.

5. Find the radius of curvature of the logarithmic spiral, of which the equation is given in Example 7, § 133.

$Ans.$ $a^{\varphi} \sqrt{[1 + (\log. a)^2]}$.

152. Problem. *To find the centre of curvature of a given curvature.*

Solution. Let x_1, y_1 be the coördinates of the centre of curvature corresponding to the point of contact x, y of the given curve, and, by B. I. § 83, $x_1 - x$ is the projection of the radius of curvature upon the axis of x.

But, by B. I. § 85, this projection is also expressed by

$$\varrho \cos. \nu = x_1 - x, \tag{615}$$

whence $x_1 = x + \varrho \cos. \nu.$ (616)

In the same way we find

$$y_1 = y + \varrho \sin. \nu \tag{617}$$

153. Corollary. If the two coördinates of the given curve are eliminated between the three equations (616), (617), and the given equation of the curve, the resulting equation, containing only the coördinates of the centre of curvature, is the equation of the curve upon which the centre of curvature is situated. This curve is called the *evolute* of the given curve, for a reason which will soon be given. The given curve is called *the involute* of its evolute.

154. Corollary. The differentials of (616) and (617) are

$$d x_1 = d x + d \varrho. \cos. \nu - \sin. \nu. \varrho \, d \nu \tag{618}$$

$$d y_1 = d y + d \varrho. \sin. \nu + \cos. \nu. \varrho \, d \nu. \tag{619}$$

But by (576), (577), and (609)

$$\sin. \nu. \varrho\, d\nu = d x \tag{620}$$

$$\cos. \nu. \varrho\, d\nu = -\, dy \tag{621}$$

which, substituted in (618) and (619), give

$$d x_1 = d\varrho. \cos. \nu \tag{622}$$

$$d y_1 = d\varrho. \sin. \nu. \tag{623}$$

155. Corollary. If τ_1 and ν_1 are the angles, which the tangent and normal to the evolute make with the axis of x, we have, by (622) and (623),

$$\text{tang. } \tau_1 = -\cot. \nu_1 = \frac{d y_1}{d x_1} = \text{tang. } \nu \tag{624}$$

whence

$$\tau_1 = \nu = \tfrac{1}{2}\pi + \tau, \ \nu_1 = \tfrac{1}{2}\pi + \nu = \pi + \tau, \tag{625}$$

so that the normal to the involute coincides with the tangent to the evolute.

156. Corollary. If s_1 is the arc of the evolute, we have, by (622) and (623),

$$d s_1 = \sqrt{[(d x_1)^2 + (d y_1)^2]} = \pm\, d\varrho, \tag{626}$$

so that the arc of the evolute increases at the same rate that the radius of curvature of the involute increases or decreases. Hence

$$\varDelta s_1 = \pm\, \varDelta \varrho. \tag{627}$$

157. Corollary. If CMM' (fig. 54.) is the involute, $C_1 M_1 M_1'$ the evolute, and if $M M_1$, $M' M_1'$ are tangent to the evolute, and consequently normal to the involute, we have

$$\varrho = MM_1, \quad \varrho' = M' M_1'$$

$$\varDelta \varrho = M' M_1' - MM_1 = \varDelta \dot{} S = M_1 M_1',$$

so that $\quad M' M_1' = MM_1 + M_1 M_1'.$

Hence if a string were wound around the evolute of such a length, that, when drawn tight at M_1 in the direction of the tangent MM_1, it would reach to M, it would, when unwound and drawn tight at M_1', reach to M', and its extremity would, in the process of un-winding, describe the involute. The names of these curves are derived from this property.

158. *Corollary.* If ϱ_1 is the radius of curvature of the evolute, we have, from (611), (625), and (626),

$$\varrho_1 = \frac{ds_1}{d\tau_1} = \pm \frac{d\varrho}{d\tau} = \pm d_c \tau \varrho. \tag{628}$$

159. *Corollary.* The evolute of the evolute is called *the second evolute,* and the evolute of the second evolute *the third evolute,* and so on.

If, then, ϱ_n is the radius of curvature of the nth evolute, and τ_n the angle, which the tangent to this evolute makes with the axis of x, (625) and (628) give

$$\varrho_n = \pm d_{c}^n \tau \varrho \tag{629}$$

$$\tau_n = \tau + \tfrac{1}{2} n \pi. \tag{630}$$

160. *Scholium.* No more natural system of coördi-nates of a curve could probably be devised than its radius of curvature, ϱ, and the angle, τ, which its di-rection makes with a given direction. A curve is

21*

readily referred to these coördinates by the equations already given ; and from its equation referred to these coördinates, the corresponding equation of either of its evolutes is readily obtained by means of (629) and (630).

161. EXAMPLES.

1. Find the evolute of the ellipse.

 Ans. The equation referred to the coördinates of § 160 is

$$\varrho = \frac{3 A^2 B^2 (A^2 - B^2) \sin. 2\tau}{(A^2 \cos.^2 \tau + B^2 \sin.^2 \tau)^{\frac{5}{2}}}.$$

2. Find the evolute of the parabola of equation of § 160.

 Ans. Its equation is

$$\varrho = \frac{6 p \sin. \tau}{\cos.^4 \tau}.$$

3. Find the nth evolute of the cycloid.

 Ans. Its equation is

$$\varrho = 4 R \cos. \tau,$$

so that it is a cycloid precisely equal to the given cycloid.

4. Find the nth evolute of the logarithmic spiral.

 Ans. It is a logarithmic spiral.

5. Find the evolute of the curve of which the equation results from eliminating φ between the two equations

$$y = R \sin. \varphi - R \varphi \cos. \varphi$$
$$x = R \cos. \varphi + R \varphi \sin. \varphi.$$

Solution. We have

$$d\,x = R\,\varphi\,d\,\varphi \text{ cos. }_\varphi$$

$$d\,y = R\,\varphi\,d\,\varphi \text{ sin. }_\varphi$$

$$\text{tang. }\tau = \text{tang. }_\varphi, \ \tau = \varphi$$

$$d\,s = R\,\tau\,d\,\tau$$

$$\varrho = R\,\tau$$

$$\varrho_1 = d_{c\cdot}\tau\,\varrho = R,$$

that is, the radius of curvature of the evolute is constant, and the evolute is therefore a circle.

CHAPTER XI.

SINGULAR POINTS.

162. Those points of a curve, which present any peculiarity as to curvature or discontinuity, are called *singular points.*

163. Whenever a function is discontinuous, the corresponding curve found, as in § 5, is also generally discontinuous.

Thus if $f. x$ is a function of x, which is imaginary for all values of x less than

$$x = a = AP \text{ (fig. 55.)};$$

for all values contained

$$\text{between } x = a' = AP' \text{ and } x = a'' = AP'',$$
$$\text{between } x = a''' = AP''' \text{ and } x = a^{\text{iv}} = AP^{\text{iv}},$$

and for all values greater than

$$x = a^{\text{v}} = AP^{\text{v}};$$

and is continuous for all values of x

$$\text{between } x = a \quad = AP \quad \text{and } x = a' = AP',$$
$$\text{between } x = a'' = AP'' \text{ and } x = a''' = AP''',$$
$$\text{and} \quad \text{between } x = a^{\text{iv}} = AP^{\text{iv}} \text{ and } x = a^{\text{v}} = AP^{\text{v}};$$

Points of stopping.

its locus is composed of the different portions MM', $M''M'''$, and $M^{iv}M^v$.

If, for instance, this function were such as to have always the same value

$$b = PM = P'N' = P''N'', \&c.$$

wherever it was not imaginary, the locus of

$$y = f. x$$

would be the portions MN', $N''N'''$, and $N^{iv}N^v$, drawn parallel to the axis of x.

164. EXAMPLES.

1. Construct the locus of the equation

$$y - b = [\log. (x - a)]^{-1}$$

in the vicinity of the point at which it stops; and find its tangent at this point.

Solution. The logarithm of a negative number is imaginary, and therefore the value of y is imaginary as long as x is less than a; but when $x = a$, we have

$$y - b = (\log. 0)^{-1} = \infty^{-1} = 0$$
$$y = b,$$

so that the point M (fig. 56.), for which

$$AP = a, \quad PM = b$$

is the point at which the curve stops. At this point we have, by § 108,

$$\text{tang. } \tau = d_c . y = - [\log. (x - a)]^{-2} (x - a)^{-1} = \infty$$
$$\tau = \tfrac{1}{2} \pi,$$

so that PM is the tangent to the curve at the point M.

Points of stopping.
---

*The remainder of the curve near the point M is constructed by finding different values of y for different values of x nearly equal to a, and drawing the curve through the points M, M', M'', &c. thus determined. The figure 56 has been constructed for the case in which

$$a = 2, \ b = 1,$$

and, for the present example, extends to

$$x = AP'' = 2.135, \ y = 0.5, \ \tau = 118° \ 26'.$$

2. Construct the locus of the equation

$$y - b = (x - a) \ [\log. \ (x - a)]^{-1}$$

in the vicinity of the point at which it stops; and find its tangent at this point.

Ans. This locus is, for the present example, represented in fig. 57, from

$$x = -\infty \ \text{to} \ x = a + 0.5 = AP'.$$

The point where the curve stops corresponds to

$$x = a = AP, \ \text{where} \ \tau = 0,$$

so that MT parallel to AX is the tangent.

3. Construct the locus of the equation

$$y - b = (x - a - 1) \ [\log. \ (x - a)]^{-1}$$

and find the tangent at the point where it stops.

Ans. This locus is represented in fig. 58. The point where it stops corresponds to

$$x = a = AP, \ \text{where} \ \tau = \tfrac{1}{2}\pi,$$

so that PM is the tangent.

Points of stopping.

4. Construct the locus of the equation

$$y - b = (x - a) \log. (x - a)$$

and find the tangent at the point where it stops.

Ans. This locus is represented in fig. 59. The point where it stops corresponds to

$$x = a = AP, \text{ where } \tau = \tfrac{1}{2}\pi,$$

so that PM is the tangent.

5. Construct the locus of the equation

$$y - b = (x - a)^2 \log. (x - a)$$

in the vicinity of the point where it stops, and find the tangent at this point.

Ans. This locus is represented in fig. 60, which, for the present example, extends

from $x = -\infty$

to $x = AP' = a + 0·223$, where $y = b - 0·075$, $\tau = 155° 57'$.

The curve stops at the point corresponding to

$$x = AP = a, \text{ where } \tau = 0,$$

so that MT, parallel to AX, is the tangent.

6. Construct the locus of the equation

$$y - b = (x - a) \left[\log. (x - a)\right]^2$$

in the vicinity of the point where it stops, and find its tangent at this point.

Ans. This locus is represented in fig. 61, which, for the present example, extends

from $x = -\infty$ to $x = AP' = a + 0·368$;

it stops at the point coresponding to

$$x = AP = a, \text{ where } \tau = \tfrac{1}{2}\pi,$$

so that PM is the tangent at this point.

7. Construct the locus of the equation

$$y - b = (x - a)^2 \,[\log. (x - a)]^2$$

in the vicinity of the point where it stops, and find the tangent at this point.

Ans. This locus is represented in fig. 62, which, for the present example, extends

$$\text{from} \quad x = -\infty, \text{ to } x = AP' = a + 0.38 \,;$$

it stops at the point corresponding to

$$x = AP = a, \text{ where } \tau = 0,$$

so that MT, parallel to AX, is the tangent.

8. Construct the locus of the equation

$$y = \log. (x + 1) + x \log. x$$

and find the tangent at the point where it stops.

Ans. This locus is represented in fig. 63; it stops at the origin where the axis of y is the tangent.

9. Construct the locus of the equation

$$y = m x \log. x + n (a - x) \log. (a - x)$$

and find the tangents at the points where it stops.

Ans. This locus stops at the points where

$$x = 0 \text{ and } x = a \,;$$

Points of stopping.

at which points the values of y are, respectively,

$$y = n \, a \, \log. \, a, \text{ and } y = m \, a \, \log. \, a;$$

and the tangents are parallel to the axis of y.

Figure 64 represents this locus when

$$a = 1, \ m = 2, \ n = 3,$$

and figure 65 represents it when

$$a = 1, \ m = 2, \ n = -3.$$

10. Construct the locus of the equation

$$y = m \, x^2 \log. \, x + n \, (a - x) \log. \, (a - x)$$

and find the tangents at the points where it stops.

Ans. This locus stops at the points where

$$x = 0 \text{ and } x = a;$$

at the first of which points

$$\text{tang. } \tau = - n \log. \, (a - x) - n,$$

and at the second

$$\tau = \tfrac{1}{2} \pi.$$

Figure 66 represents this curve when

$$a = 1, \ m = 2, \ n = 3;$$

and figure 67 represents it when

$$a = 1, \ m = 2, \ n = -3.$$

11. Construct the locus of the equation

$$y = f_1 . x \, (f. \, x)^n \log. f. \, x$$

in which $f. \, x$ is a given function of x; and find the points where it stops.

22

Ans. It stops when $f.x$ becomes imaginary, or when it becomes negative.

Figure 68 represents this locus when

$$f_1.x = n = 1, \quad f.x = x^2 - x,$$

in which case it extends

from $\quad x = -\infty$ to $x = 0,$

where it stops, and extends again

from $\quad x = 1 = AP$ to $x = \infty.$

The tangents at each of the points where it stops is parallel to the axis of x.

Figure 69 represents this locus when

$$n = 1, \ f_1.x = x, \ f.x = x^2 - x,$$

so that the points of stopping are the same as in figure 68. But the tangent at the point A is the axis of x, while that at the point P is parallel to the axis of y.

Figure 70 represents this locus when

$$n = 2, \quad f_1.x = 1, \quad f.x = x^2 - x,$$

so that the points of stopping are the same as in figure 68; but the tangent at each point is the axis of x.

Figure 71 represents this locus when

$$f_1.x = n = 1, \quad f.x = x - x^2,$$

so that the points of stopping and the tangents are the same as in figure 68; but the curve extends from one point to the other.

Figure 72 represents this locus when

$$n = 1, \quad f_1.x = x, \quad f.x = x - x^2,$$

Points of stopping.

so that the points of stopping and the tangents are the same as in figure 69; but the curve extends from one point to the other.

Figure 73 represents this locus when

$$n = 2, \quad f_1 . x = 1, \quad f . x = x - x^2,$$

so that the points of stopping and the tangents are the same as in figure 70; but the curve extends from one point to the other.

Figure 74 represents this locus when

$$n = 1, \, f_1 . x = (10 \, x + 1)^{-1} \cdot$$
$$f . x = x \, (x - 1) \, (x - 2) \, (x - 4) \, (x - 5)$$

in which case it extends

from $x = 0$ to $x = 1 = AP_1$ where it stops,

from $x = 2 = AP_2$ to $x = 4 = AP_4$ where it stops,

from $x = 5 = AP_5$ to $x = \infty$.

The tangent at each point where it stops is parallel to the axis of y.

Figure 75 represents this locus when

$$n = 1, \, f_1 . x = (10 \, x + 1)^{-1} \cdot$$
$$f . x = - x \, (x - 1) \, (x - 2) \, (x - 4) \, (x - 5),$$

in which case it extends

from $x = -\infty$ to $x = 0$,   where it stops,

from $x = 1 = AP_1$ to $x \equiv 2 = AP_2$, where it stops,

from $x = 4 = AP_4$ to $x = 5 = AP_5$, where it stops;

the points of stopping and the tangents are the same as in figure 74.

Figure 76 represents this locus when

$$n = 1, f_1 . x = \tfrac{1}{2}(x+1) x^2 (x-2)(x-3)(x-4)(2x+3)^{-2}$$
$$f. x = x (x-1)(x-2)(x-4)(x-5);$$

in which case it extends as in figure 74, and the tangents at the points P_1 and P_5 are parallel to the axis of y, but the axis of x is the tangent at the points A, P_2, and P_4.

Figure 77 represents this locus when

$$n = 1, f_1 . x = 4 (x+1) x^2 (x-2)(x-3)(x-4)(10x+1)^{-4}$$
$$f. x = - x (x-1)(x-2)(x-4)(x-5),$$

in which case, it extends as in figure 75, and the tangents are as in figure 76.

Figure 78 reprents this locus when

$$n = 1, \quad f_1 . x = (10 x + 1)^{-1}$$
$$f. x = x (x-1)(x-2)(x-3)(x-4)(x-5),$$

in which case it extends

from $x = - \infty$ to $x = 0$, where it stops,

from $x = 1 = AP_1$ to $x = 2 = AP_2$, where it stops,

from $x = 3 = AP_3$ to $x = 4 = AP_4$, where it stops,

from $x = 5 = AP_5$ to $x = \infty$;

the tangent at each point where it stops is parallel to the axis of y.

Figure 79 represents this locus when

$$n = 1, \quad f_1 . x = \tfrac{1}{10} (10 x + 1)^{-1}$$
$$f. x = - x (x-1)(x-2)(x-3)(x-4)(x-5),$$

in which case it extends

Points of stopping.

from $x = 0$ to $x = 1 = AP_1$, where it stops,

from $x = 2 = AP_2$ to $x = 3 = AP_3$, where it stops,

from $x = 4 = AP_4$ to $x = 5 = AP_5$, where it stops ;

the points of stopping and the tangents are the same as in figure 78.

Figure 80 represents this locus when

$$n = 1, \ f_1 \cdot x = \tfrac{1}{2}(x+1)x^2(x-2)(x-3)(x-4)(2x+3)^{-2}$$
$$f \cdot x = x(x-1)(x-2)(x-3)(x-4)(x-5),$$

in which case it extends as in figure 78, and the tangents at the points P_1 and P_5 are parallel to the axis of y; but the axis of x is the tangent at the points A, P_2, P_3, and P_4.

Figure 81 represents this locus when

$$n = 1, \ f_1 \cdot x = 4(x+1)x^2(x-2)(x-3)(x-4)(10x+1)^{-4}$$
$$f \cdot x = -x(x-1)(x-2)(x-3)(x-4)(x-5),$$

in which case it extends as in figure 79, and the tangents are as in figure 80.

Figure 82 represents this locus when

$$n = 1, \ f_1 \cdot x = \tfrac{1}{4}x$$
$$f \cdot x = -6x(x-1)^2(x-2),$$

in which case it extends from

$$x = 0 \text{ to } x = 2 = AP_2.$$

Figure 83 represents this locus when

$$n = 1, \ f_1 \, x = \tfrac{1}{4}x(x-1)^{-1}$$
$$f \cdot x = -6x(x-1)^2(x-2),$$

in which case it extends as in figure 82.

22*

Points of stopping.

Figure 84 represennts this locus when

$$n = 1, \quad f_1. x = 1$$
$$f. x = x + \sqrt{x},$$

in which case the portion AM of the curve, which corresponds to the positive value of the radical, extends

from $\quad x = 0$ to $x = \infty$;

the portion $P_1 M_1$, which corresponds to the negative value of the radical extends

from $\quad x = 1 = AP_1$ to $x = \infty$.

Figure 85 represents this locus when

$$n = 2, \quad f_1. x = 1$$
$$f. x = x + \sqrt{x},$$

in which case the portions extend as in figure 84.

Figure 86 represents this locus when

$$\_ \quad n = 0, \quad f_1. x = (x^2 - x)$$
$$f. x = x + \sqrt{x},$$

in which case the portions extend as in figure 84.

Figure 87 represents this locus when

$$n = 1, \quad f_1. x = \log. f. x$$
$$f. x = x + \sqrt{x},$$

in which case the portions extend as in figure 84.

Figure 88 represents this locus when

$$n = 1, \quad f_1. x = 1$$
$$f. x = (x + \sqrt{x})^2,$$

Points of stopping.

in which case each portion extends from

$$x = 0 \text{ to } x = \infty.$$

Figure 89 represents this locus when

$$n = 1, \quad f_1 . x = (x + \sqrt{x})^{-1}$$
$$f . x = (x + \sqrt{x})^2$$

in which case the portions extend as in figure 88.

Figure 90 represents this locus when

$$n = 1, \quad f_1 . x = \log . f . x$$
$$f . x = (x + \sqrt{x})^2,$$

in which case the portions extend as in figure 88.

Figure 91 represents this locus when

$$n = 0, \quad f_1 . x = (x^2 - x)^2 \log . f . x$$
$$f . x = (x + \sqrt{x})^2,$$

in which case the portions extend as in figure 88.

Figures 92 – 99 represent this locus when

$$f_1 . x = n = 1, \quad f . x = a + \sqrt{(4 - x^2)},$$

in which case it stops at the values

$$x = -AP' = -\sqrt{(4 - a^2)} \text{ and } x = AP'' = \sqrt{(4 - a^2)}.$$

The tangents at the points P' and P'' are parallel to the axis of y.

In figure 92, $a = -1.5$.

In figure 93, $a = -1$.

In figure 94, $a = -0.5$.

In figure 95, $a = 0$.

Points of stopping.

In figure 96, $a = 0.5$.

In figure 97, $a = 1$.

In figure 98, $a = 1.5$.

In figure 99, $a = 2$.

12. Construct the locus of the equation

$$y = f. x + m f_1. x$$

when m is infinitely small and $f_1. x$ a junction, which is not infinite while x is finite, but is alternately real and imaginary.

Solution. Since m is infinitely small, the part $m f_1. x$ may be neglected when $f_1. x$ is real, but when $f_1. x$ is imaginary, the value of y is imaginary, so that if figure 100 is the locus of equation

$$y = f. x ;$$

the same figure, with the dotted parts omitted, which correspond to the imaginary values of $f_1. x$, represents the locus of the given equation.

Thus, the locus of the equation

$$y = \sqrt{(R^2 - x^2)} + 0 \cdot 00000000001 \times x^n \log. x$$

differs insensibly from the semicircle BCB' (fig. 101.), of which R is the radius. But it must be remarked, that, when n is zero, the curve is suddenly turned into the form of a hook at the points B and B', so as to become tangent to the axis of y, assuming a form similar to that of the dotted line, but of indefinitely less extent.

13. Construct the locus of the equation

$$r = f. \varphi + m f_1. \varphi$$

expressed in polar coördinates, and in which m is infinitely small, and $f_1 \cdot \varphi$ finite when real.

Solution. If $MM' M''M'''$ &c. fig. (102.) represent the locus of

$$r = f. \varphi,$$

the same curve with the dotted parts omitted represents the given curve, the dotted parts corresponding to the imaginary values of $f_1 \cdot \varphi$.

Thus, if $\qquad f. \varphi = R,$

the curve consists of several successive arcs of the same circle.

165. A *conjugate* point is one separated entirely from the rest of the curve, but included in the same algebraic equation.

A conjugate point is indicated algebraically by the condition that coördinates of this point are real, while the coördinates of no adjacent point are so.

166. EXAMPLES.

1. Construct the locus of the equation

$$y = f. x + m f_1 . x,$$

in which m is imaginary and $f_1 . x$ real.

Solution. If the curve (fig. 100.) is the locus of

$$y = f. x,$$

and if M, M', &c. are the points which correspond to the abscissas, for which

$$f_1 . x = 0 \, ;$$

the locus of the given equation consists of the series of conjugate points M, M', &c., without any continuous curve.

If $\qquad f . x = a x + b,$

all these points are upon the same straight line.

2. Construct the locus of the equation

$$(y - f . x)^2 + (y - f_1 . x)^2 = 0.$$

Solution. The sum of two squares cannot be zero unless each square is zero; so that the given equation is equivalent to the two equations

$$y - f . x = 0, \qquad y - f_1 . x = 0 \, ;$$

that is, the coördinates of all the points of the required locus satisfy these two equations.

If, then, $APP'P''P'''P^{\mathrm{iv}}P^{\mathrm{v}}$ &c. (fig. 103.) is the locus of the equation

$$y = f . x,$$

and if $AP_1 \, P' \, P_1' \, P'' \, P_1'' \, P''' \, P^{\mathrm{iv}} \, P_1^{\mathrm{iv}}$ &c. is the locus of the equation

$$y = f_1 . x,$$

the required locus is the series of conjugate points A, P', P'', P^{iv}, &c., in which these curves intersect.

Thus the locus of the equation

$$(x - a)^2 + (y - b)^2 = 0$$

is the single conjugate point of which the coördinates are a and b.

3. Find the conjugate points of the locus of the equation

$$y = f.x + f_1.x\,f_2.x,$$

in which $f_1.x$ and $f_2.x$ are sometimes imaginary.

Solution. If $f_1.x$ is imaginary for values of x between a and b, and, if $f_2.x$ vanishes for one or more of the values of x contained between a and b; the given equation is reduced for these values of $f_2.x$ to

$$y = f.x,$$

so that the corresponding points of the curve are conjugate points situated upon the locus of the equation

$$y = f.x.$$

In the same way, those points of this locus are conjugate which correspond to values of x, for which $f_1.x$ vanishes, while $f_2.x$ is imaginary.

Thus the point P', for which $x = -1$ is a conjugate point upon the axis of x in the curve of figure 76.

This locus is represented in figure 104, when

$$f.x = \sqrt{(4 - x^2)},\ f_1.x = \sqrt{(1 - x^2)},\ f_2.x = x\log.x - 1.$$

It has four conjugate points, M', M_1', M'', M_1'', situated upon the circle of which the origin is the centre, and of which the radius is

$$AP = -2.$$

The common abscissa of two of these points is

$$x = -AP' = -1,$$

Branch.	Multiple points.	Cusp.

and of the other two

$$x = AP'' = 1{\cdot}763.$$

4. Construct the locus of the polar equation of example 13, § 164, when m is imaginary.

Ans. It represents a series of conjugate points, upon the curve of which the equation is

$$r = f. \varphi.$$

These points correspond to the values of φ, which satisfy the equation

$$f_1. \varphi = 0.$$

When $f. \varphi = R,$

the points are all situated upon the circumference of which R is the radius, and the origin the centre.

167. A *branch* of a curve is a continuous portion of it, which extends from one point of discontinuity to another.

When a branch returns into itself, so that its commencement is the continued curve of its end, it is called an *oval*.

168. A point through which the curve passes more than once, or at which two or more branches terminate, is called a *multiple point*.

A multiple point at which two or more branches stop, and have the same tangent, is called a *cusp*. If a branch begins and ends at a point, having but one tangent at this point, without being continuous, this point is also a cusp.

A cusp is said to be of *the first kind*, when the two branches at the point of contact lie upon opposite sides of the tangent, as at *M* (fig. 106.) ; but if the two branches lie upon the same side of the tangent, as at *M* (fig. 107), the cusp is said to be of *the second kind*.

169. In the algebraic consideration of curves, they are naturally divided into *portions*, according to the number of ordinates which correspond to the same abscissa ; or of radii vectores, which correspond to the same angle.

The *algebraic portions* of a curve are not to be confounded with the *geometric branches;* for the same portion may consist of several branches, or several different portions may be united into one branch.

Thus the cycloid consists of but one portion, but of an infinite number of branches ; whereas the circle, the ellipse, and the parabola consist of two portions, but only of one branch ; and though the hyperbola consists of two portions and two branches, yet half of each portion belongs to each branch.

170. *Problem.* *To find the cusps of a given curve.*

Solution. 1. If a portion $M'MM''$ (fig. 105.) of a curve, whose equation is expressed in rectangular coördinates, has a cusp at a point M, it is evident that the tangent TM at this point must be perpendicular to the axis of x. For if it were not so, as in figure 106, there would, for the abscissa AP' very near to AP, be the two ordinates $P'M'$ and $P'M'_1$, so that MM' and MM'_1 would be two different portions of the curve, and not the same portion, as we here suppose.

23

Moreover, the tangents $M'T'$ and $M''T'''$, which are infinitely near to MT must evidently be inclined to the axis of x, one by an acute angle and the other by an obtuse angle, so that

$$\text{tang. } \tau = d_c.y$$

must change its sign at the point M, by passing through infinity if the point M is a cusp, formed by two branches of the same portion of the curve ; and such a cusp is necessarily of the first kind.

2. If a cusp is formed at the meeting of two branches of different portions, as at M (figs. 106 and 107) and if the common tangent MT is not perpendicular to the axis of x ; the ordinates for both portions, which correspond to the abscissas AP and AP', one of which is greater and the other less than AP, must be imaginary for one of these abscissas, and real for the other. The cusp is of the first kind, as in figure 106, if the value of τ is greater than MTX upon one branch, and less than MTX upon the other branch ; but it is of the second kind, as in figure 107, if the value of τ is greater or less than MTX upon both branches.

But if the common tangent is perpendicular to the axis of x at M (figs. 108 and 109), the ordinates for the two portions must be both increasing, or both increasing in proceeding from M. The cusp is of the first kind, as in figure 108, when it is the end of one branch and the beginning of the other ; but it

Branches.	Oval.

is of the second when it is the end or the beginning of both branches, as in fig. 109.

171. *Problem.   To find the points where two portions of a curve unite in the same branch.*

Solution.   The point M (figs. 110 *and* 111) *is one of the required points, if the two portions MM' and MM_1' have a common tangent at this point, while the point is not a cusp, but ·merely a point where both the portions stop.*

172. *Corollary.   The portions $M\,M'\,M_2$ and $MM_1'M_2$ (fig.* 112) *compose an oval, if at their two extremities M and M_2 they unite in a continuous curve and have no point of discontinuity between their extremities.*

173. When the curve is expressed in polar coördinates, the analytic portion depends upon the number of radii vectores which correspond to the same angle. But it must not be overlooked that the same direction is determined by angles which differ by any entire multiple of four right angles, so that a curve like one of the spirals of B. 1, § 98, may consist of but one portion, although there are an infinite number of radii vectores in each direction.

Multiple points are obtained in any portion, when the same value of the radius corresponds to two or more angles, which differ by any entire multiple of four right angles.

174. Examples.

1. Find the cusps of the cycloid.

Solution. The cycloid obviously consists of but a single portion. If there is a cusp, the tangent at it must, then, as in § 170, be perpendicular to the axis of x; that is, we must have, by § 131, example 5,

$$\text{cotan. } \tfrac{1}{2}\, \theta = \infty,$$

which gives

$$\tfrac{1}{2} \theta = n\, \pi, \; \theta = 2\, n\, \pi,$$

in which n is an integer.

But this value of θ gives, by (131),

$$y = 0,$$

and since we can never have

$$\cos. \, \theta > 1,$$

the value of y is never negative, so that there is a cusp at each of the points where

$$y = 0.$$

2. Find the branches of the locus of Example 1, § 164.

Ans. It consists of two branches, one of which, MM' (fig. 56.), begins with

$$x = a, \quad y = b,$$

and extends to

$$x = a + 1, \quad y = -\infty.$$

The second $M_1 M_1'$ begins with

$$x = a + 1, \quad y = \infty,$$

and extends to
$$x = \infty, \quad y = b.$$

3. Find the branches of the locus of Example 2, § 164.

Ans. It consists of two branches, one of which MM' (fig. 57.), begins with
$$x = a, \quad y = b,$$
and extends to
$$x = a + 1, \quad y = -\infty.$$

The second $M_1 M_1'$, begins with
$$x = a + 1, \quad y = \infty,$$
and extends to
$$x = \infty, \quad y = \infty.$$

The least value of y in this second branch is found by § 113, to be
$$y = P_1' M_1' = 2.718, \text{ corresponding to } x = AP_1' = 2.718.$$

4. Find the multiple point of the locus of Example 11, § 164, when
$$n = 0, \quad f_1 . x = (x^2 - x)$$
$$f . x = x + \sqrt{x}.$$

Ans. There is a multiple point when
$$x = 1 = AP_2, \quad y = 0,$$
at which point the portion corresponding to the negative **value** of the radical begins, its tangent being $P_2 T_2$ (fig. 86.), drawn parallel to the axis of y, and the portion corresponding to the positive value of the radical passes through the same point, its tangent being $P_2 T_2'$, so drawn that
$$P_2' T_2 X = 34° 44'.$$

23*

5. Find the multiple points of the locus of **Example 11**, § 164, when

$$n = 1, \quad f_1 . x = \log . f. x$$
$$f. x = \tfrac{1}{2} (x + \sqrt{x}).$$

Ans. There are two multiple points; one is at P_1 (fig. 113.) where

$$x = 1, \quad y = 0 \, ;$$

at which point the branch corresponding to the negative value of the radical begins, while the other branch passes through it; $P_1 T_1$ is the tangent to the former branch, and the axis of x is tangent to the latter branch. The other multiple point is at M_2, where

$$x = 1\cdot 96, \quad y = 0\cdot 45 \, ;$$

the value of τ for the former branch is

$$\tau = 149° \ 15',$$

and that for the latter branch is

$$\tau = 60° \ 35'.$$

6. Find the cusp and the other multiple point of the locus of Example 11, § 164, when

$$n = 1, f_1 . x = 1, f. x = (x + \sqrt{x})^2.$$

Ans. The origin A (fig. 88.) is a cusp of the second kind, and the axis of x is the tangent at this point.

The other multiple point M_1 corresponds to

$$x = 0\cdot 328, \quad y = 0\cdot 169.$$

The values of τ at this point are

$$\tau = 69° \ 29', \text{ and } \tau = 6° \ 30'.$$

Branches and multiple points.

7. Find the branches and the multiple point of the locus of Example 11, § 164, when

$$n = 1, \quad f_1 . x = (x + \sqrt{x})^{-1}, \quad f. x = (x + \sqrt{x})^2.$$

Ans. The curve consists of but one branch, for the two portions unite in one branch at the origin A (fig. 89.)

The multiple point corresponds to

$$x = 0\cdot544, \quad y = 0\cdot634,$$

at which point the two values of τ are

$$\tau = 76° \ 35', \quad \tau = 124° \ 13'.$$

8. Find the multiple points of the locus of Example 11, § 164, when

$$n = 1, \quad f_1 . x = \log. f. x$$
$$f. x = (x + \sqrt{x})^2.$$

Ans. The origin (fig. 90.) is a cusp of the second kind, the tangent at this point being the axis of x.

M_1 is a multiple point corresponding to

$$x = 0\cdot142, \quad y = 0\cdot465,$$

at which point the two values of τ are

$$\tau = 114° \ 37', \quad \tau = 21° \ 40'.$$

M_2 is a multiple point corresponding to

$$x = 0\cdot544, \quad y = 0\cdot402,$$

at which point the two values of τ are

$$\tau = 53° \ 17', \quad \tau = 172° \ 29'.$$

9. Find the multiple points of the locus of Example 11, § 164, when

$$n = 0, \quad f_1 x = (x^2 - x)^2 \log. f. x$$
$$f. x = (x + \sqrt{x})^2.$$

Ans. The origin (fig. 91.) is a cusp of the second kind, the tangent at this point being the axis of x.

M_1 is a multiple point corresponding to

$$x = 1, \quad y = 0,$$

at which the axis of x is the common tangent to the two branches of the curve, and the contact of the two branches is of the second order.

10. Has the locus of Example 11, § 164, any cusp, when

$$n = 1, f_1. x = \tfrac{1}{4} x (x-1)^{-1}, f. x = -6x(x-1)^2 (x-2) ?$$

Ans. It has none.

11. Find the multiple point of the locus of Example 11, § 164, when

$$f_1. x = n = 1, \quad f. x = a + \sqrt{(4 - x^2)}.$$

Ans. When a is zero, or negative, there is no multiple point.

When a is positive, and less than

$$e^{-1} = [2 \cdot 71828]^{-1} = 0 \cdot 3679,$$

the curve consists of a single branch without any multiple point.

When $a = e^{-1} = 0 \cdot 3679$

the curve consists of three branches, as in (fig. 114), and has two cusps of the second kind, correspondsng to

$$x = \pm 2, \quad y = -a = -0 \cdot 3679.$$

Branches and multiple points.

When a is greater than e^{-1} and less than $\frac{1}{2}$, the curve consists of one branch with two multiples, as in figure 115, where

$$a = 0{\cdot}4,$$

and the two multiple points correspond to

$$x = \pm \ 1{\cdot}984, \quad y = - \ 0{\cdot}275.$$

The values of τ at one point are

$$\tau = 102^\circ \ 38', \text{ and } \tau = 97^\circ \ 45',$$

at the other

$$\tau = 77^\circ \ 22', \text{ and } \tau = 82^\circ \ 15'.$$

When $$a = \tfrac{1}{2} = 0{\cdot}5,$$

the curve (fig. 96.) has two multiple points at the beginning and end of its branch, corresponding to

$$x = \pm \ 1{\cdot}937, \quad y = 0.$$

The values of τ are

$$\tau = 90^\circ \ , \ \ \tau = 90^\circ \pm 45^\circ \ .$$

When a is greater than $\frac{1}{2}$ and less than 2, the curve consists of a single branch, with no multiple points.

When $$a = 2,$$

the curve (fig. 99.) is an oval.

12. Construct the locus of Example 11, § 164, when

$$f_1 . x = n = 1, \quad f . x = a + \sqrt{(a^2 - x^2)}.$$

Ans. Where a is greater than $\frac{1}{2}$, the curve is an oval, as in figure 99, where

$$a = 2.$$

When $$a = \tfrac{1}{2} = 0{\cdot}5,$$

the curve (fig. 116.) consists of a single continuous branch, which returns into itself; and it has a multiple point at the origin, where the curve has a contact with itself, the common tangent being the axis of x.

When a is less than $\frac{1}{2}$ and greater than

$$e^{-1} = 0.3679,$$

the curve consists of a single continuous branch, which returns into itself; and has two multiple points, as in figure 117, where

$$a = 0.4,$$

the two multiple points correspond to

$$x = \pm\, 0.31, \quad y = -\, 0.27.$$

The values of τ at one point are

$$\tau = 144°\ 53' \quad \tau = 131°\ 41',$$

and at the other

$$\tau = 35°\ 7' \quad \tau = 45°\ 17'.$$

When $\qquad a = e^{-1} = 0.3679,$

the curve (fig. 118.) consists of two branches and two cusps of the second kind, corresponding to

$$x = a, \quad y = -\, a.$$

When a is less than e^{-1}, the curve is an oval, as in figure 119, where

$$a = 0.2.$$

13. Construct the locus of the equations of Example 5, § 161, and find its cusp.

Ans. This locus is represented in figure 120. Its cusp is of the first kind, and corresponds to

Branches and multiple points.

$$\varphi = 0, \quad x = 1, \quad y = 0,$$

where the axis of x is the tangent.

14. Construct the locus of the equation

$$y^2 = x^3,$$

and find its cusp.

Ans. This locus is represented in figure 121. Its cusp is of the first kind, and is the origin, where the axis of x is the tangent.

15. Construct the locus of the equation

$$y^2 = x^4 - x^3,$$

and find its cusp.

Ans. This curve (fig. 122.) consists of three branches, two of which extend from

$$x = -\infty \quad \text{to} \quad x = 0,$$

where there is a cusp of the first kind.

The third branch extends from

$$x = 1 \quad \text{to} \quad x = \infty.$$

16. Construct the locus of the equation

$$y^2 = x^3 - x^4,$$

and find its cusp.

Ans. This locus (fig. 123.) consists of a single branch, which has a cusp of the first kind at the origin.

17. Construct the locus of the equation

$$y^2 = x^4 - x^2,$$

and find its branches.

Ans. This locus (fig. 124.) consists of two branches, one of which extends from

$$x = -\infty \text{ to } x = -1,$$

and the other from

$$x = 1 \text{ to } x = \infty,$$

and a conjugate point, which is the origin.

18. Construct the locus of the equation

$$y^2 = x^2 - x^4,$$

and find its multiple point.

Ans. This locus (fig. 125) consists of one branch, which returns into itself, and has a multiple point at the origin, where the values of τ are

$$\tau = \pm 45°.$$

19. Construct the locus of the equation

$$y^2 = x^4 - x^6,$$

and find its multiple point.

Ans. This locus (fig. 126.) consists of a single branch, which returns into itself, and has a multiple point at the origin, where it has a contact with itself. The tangent at the origin is the axis of x.

20. Construct the locus of the equation

$$y^2 = x^3 - x.$$

Ans. This locus (fig. 127.) consists of an oval, which extends from

Multiple points.

$$x = -1 \text{ to } x = 0,$$

and a branch which extends from

$$x = 1 \text{ to } x = \infty.$$

21. Construct the locus of the equation

$$y^2 = x^5 - x^3,$$

and find its cusp.

Ans. This locus (fig. 128.) consists of a branch, which extends from

$$x = -1 \text{ to } x = 0,$$

where there is a cusp, and a branch, which extends from

$$x = 1 \text{ to } x = \infty.$$

22. Construct the locus of the equation

$$y^2 = x^2 (1 - x^2)^3,$$

and find its multiple points.

Ans. This locus (fig. 129.) consists of two branches, which extend from

$$x = -1 \text{ to } x = 1.$$

They cross at the origin, where the values of τ are

$$\tau = \pm 45°,$$

and there are two cusps of the first kind, corresponding to

$$x = \pm 1.$$

23. Construct the locus of the equation

$$y^2 = x^4 (1 - x^2)^3,$$

and find its multiple points.

24

Ans. This locus (fig. 130.) consists of two branches, which extend from

$$x = -1 \text{ to } x = 1.$$

There are two cusps corresponding to the two values of x, and the origin is also a multiple point, where the two branches are in contact.

24. Construct the locus of the equation

$$y^2 = (2 - x^2)\,(1 - x^2)\,(\tfrac{1}{2} - x^2),$$

and find its branches.

Ans. This locus (fig. 131.) consists of a succession of three ovals, which extend respectively

from $\quad x = -\sqrt{2} \text{ to } x = -1$

from $\quad x = -\sqrt{\tfrac{1}{2}} \text{ to } x = \sqrt{\tfrac{1}{2}}$

from $\quad x \quad = \quad 1 \text{ to } x = \sqrt{2}.$

25. Construct the locus of the polar equation

$$r = a + \sin. m\ \varphi,$$

and find its multiple points and branches.

Ans. If m is an integer and a greater than 1, this locus is oval, as in figure 133, where

$$a = 2,\ m = 3,$$

If m is a fraction and a greater than 1, this locus is a curve, which returns into itself after as many revolutions of the radius vector as there are integers in the denominator of m.

Thus, in (fig. 134.),

$$a = 2,\quad m = \tfrac{1}{2},$$

there is a multiple point corresponding to

$$\varphi = 0 \text{ or } = 360°, \quad r = 2,$$

at which point the values of ε and τ are

$$\varepsilon = \tau = 75° \ 58', \text{ and } \varepsilon = \tau = 104° \ 2'.$$

In (fig. 135.) $a = 2, \quad m = \frac{3}{2},$

there are three multiple points corresponding to

$$\varphi = 0° \text{ or } = 360°, \quad \varphi = 120° \text{ or } = 480°, \quad \varphi = 240° \text{ or } = 600°,$$

at each of which points the values of r and ε are

$$\tau = 2, \ \varepsilon = 53° \ 8', \text{ and } \varepsilon = 126° \ 52'.$$

In (fig. 136.) $a = 2, \ m = \frac{1}{3},$

there are two multiple points, one of which corresponds to

$$\varphi = 90° \text{ or } = 450°, \ r = 2.5$$

$$\varepsilon = 83° \ 25' \text{ and } \varepsilon = 96° \ 35',$$

and the other to $\varphi = 630° \text{ or } = 990°, \ r = 1·5.$

$$\varepsilon = 79° \ 7' \text{ and } \varepsilon = 100° \ 53'.$$

In (fig. 137.) $a = 2, \ m = \frac{2}{3},$

there are four multiple points; at two of these points we have

$$\varphi = 45° \text{ or } = 765°, \ \varphi = 225° \text{ or } = 585°,$$

and at each of these points

$$r = 2.5, \ \varepsilon = 77°, \text{ and } \varepsilon = 103°;$$

at the other two points we have

$$\varphi = 315° \text{ or } = 1035°, \ \varphi = 495° \text{ or } = 855°,$$

and at each of these points

$$r = 1.5, \; \epsilon = 68° \; 57', \text{ and } \epsilon = 111° \; 3'.$$

In (fig. 138) $\qquad a = 2, \qquad m = \frac{1}{4},$

there are three multiple points; one of which corresponds to

$$\varphi = 0° \text{ or } = 720°, \; r = 2$$

$$\epsilon = 82° \; 53' \text{ and } \epsilon = 97° \; 7' ;$$

the second corresponds to

$$\varphi = 180° \text{ or } = 540°, \; r = 2.707$$

$$\epsilon = 86° \; 16' \text{ and } \epsilon = 93° \; 44' ;$$

the third corresponds to

$$\varphi = 900° \text{ or } = 1260°, \; r = 1.293$$

$$\epsilon = 82° \; 13' \text{ and } \epsilon = 97° \; 47'.$$

In (fig. 139.) $\qquad a = 2, \; m = \frac{3}{4},$

there are nine multiple points; three of which correspond to

$$\varphi = 0° \text{ or } = 720°, \; \varphi = 240° \text{ or } = 960°, \; \varphi = 480° \text{ or } = 1200°,$$

and at each of these points we have

$$r = 2, \; \epsilon = 69° \; 27', \text{ and } \epsilon = 110° \; 33' ;$$

three correspond to

$$\varphi = 60° \text{ or } = 1140°, \; \varphi = 180° \text{ or } = 540°, \; \varphi = 660° \text{ or } = 1020°,$$

and at each of these points, we have

$$r = 2.707, \; \epsilon = 78° \; 55', \text{ and } \epsilon = 101° \; 5' ;$$

three correspond to

Multiple points.

$\varphi=300°$ or $= 1380°$, $\varphi= 420°$ or $= 780°$, $\varphi=900°$ or $= 1260°$

and at each of these points we have

$$r = 1.293, \quad \iota = 67° \ 42', \text{ and } \iota = 112° \ 18'.$$

If $a = 1$ and m an uneven integer, or a fraction whose numerator is uneven, the origin is a multiple point consisting of the union of as many cusps as there are in the integer m, or the numerator of the fraction m.

Thus, in (fig. 140.) $\qquad a = 1, \ m = 1,$

the origin is a cusp, and the tangent is the axis of y.

In (fig. 141.) $\qquad a = 1, \ m = 3,$

the origin is an union of three cusps, and the three values of τ at this point are

$$\tau = 90°, \ \tau = 210°, \ \tau = 330°.$$

In (fig. 142.) $\qquad a = 1, \ m = 5,$

the origin is an union of five cusps, and the five values of τ at this point are

$$\tau = 54°, \ \tau = 126°, \ \tau = 198°,$$

$$\tau = 270°, \ \tau = 342°.$$

In (fig. 143.) $\qquad a = 1, \ m = \frac{1}{2},$

the origin is a cusp, and the tangent at this point is the axis of x. There is a multiple point corresponding to

$$\varphi = 0° \text{ or } = 360°, \ r = 1$$

$$\iota = 63° \ 26' \text{ and } \iota = 116° \ 34'.$$

In (fig. 144.) $\qquad a = 1, \ m = \frac{3}{2},$

24*

the origin as an union of three cusps, and the three values of τ at this point are

$$\tau = 60°, \; \tau = 180°, \; \tau = 300°.$$

There are three other multiple points corresponding to

$$\varphi = 0° \text{ or } = 360°, \; \varphi = 120° \text{ or } = 480°, \; \varphi = 240° \text{ or } = 600°,$$

at each of which points we have

$$r = 1, \; \iota = 33° \; 41', \text{ and } \varepsilon = 146° \; 19'.$$

In (fig. 145.) $a = 1, \; m = \frac{1}{3}$,

the origin is a cusp, and the tangent at this point is perpendicular to the axis.

There are two other multiple points corresponding to

$$\varphi = 90°, \text{ or } = 450°, \; \varphi = 630° \text{ or } = 990°,$$

at each of which points we have

$$r = 1, \; \iota = 71° \; 34', \text{ and } \iota = 108° \; 26'.$$

In (fig. 146.) $a = 1, \; m = \frac{5}{3}$,

the origin is an union of five cusps, and the five values of τ at this point are

$$\tau = 18°, \; \tau = 90°, \; \tau = 162°, \; \tau = 234°, \; \tau = 306°.$$

There are ten other multiple points; five of these points correspond to

$$\varphi = 90° \text{ or } = 450°, \; \varphi = 306° \text{ or } = 666°, \; \varphi = 522° \text{ or } = 882°$$

$$\varphi = 18° \text{ or } = 738°, \; \varphi = 234° \text{ or } = 954°;$$

for each of these points

$$r = 1.5, \; \iota = 46° \; 6', \text{ and } \iota = 133° \; 54',$$

Multiple points.

the other five points correspond to

$\varphi = 198°$ or $= 558°$ $\varphi = 414°$ or $= 774°$, $\varphi = 630°$ or $= 990°$,

$\varphi = 126°$ or $= 846°$, $\varphi = 342°$ or $= 1102°$.

for each of these points

$$r = 0.5, \; \varepsilon = 19° \, 6' \text{ and } \varepsilon = 160° \, 54'.$$

In (fig. 147.) $a = 1, \; m = \frac{1}{5},$

the origin is a cusp, and the tangent at this point is perpendicular to the axis.

There are four other multiple points corresponding to

$\varphi = 270°$ or $= \;\; 630°, r = 9.045, \varepsilon = 86° \, 16'$ and $\varepsilon = \;\; 93° \, 44'$

$\varphi = 1170°$ or $= 1530°, r = 0.955, \varepsilon = 58° \, 23'$ and $\varepsilon = 121° \, 37'$

$\varphi = \;\; 90°$ or $= \;\; 810°, r = 6.545, \varepsilon = 81° \, 39'$ and $\varepsilon = \;\; 98° \, 21'$

$\varphi = 990°$ or $= 1710°, r = 0.345, \varepsilon = 19° \, 58'$ and $\varepsilon = 160° \, 2'$

In (fig. 148.) $a = 1, \; m = \frac{3}{5},$

the origin is a union of three cusps, at which point the values of τ are

$$\tau = 90°, \; \tau = 210°, \; \tau = 330°.$$

There are twelve other multiple points; three of these points correspond to

$\varphi = 90°$ or $= 810°$, $\varphi = 690°$ or $= 1410°$, $\varphi = 210°$ or $= 1290°$

for each of these points

$$r = 1.809, \; \varepsilon = 78° \, 58' \text{ and } \varepsilon = 101° \, 2';$$

three points correspond to

Multiple points.

$\varphi = 270°$ or $= 630°$, $\varphi = 870°$ or $= 1230°$, $\varphi = 30°$ or $= 1470°$

for each of these points

$$r = 1.309, \quad \epsilon = 66° \, 27', \text{ and } \epsilon = 113° \, 33';$$

three points correspond to

$\varphi = 570°$ or $= 930°$, $\varphi = 1170°$ or $1530°$, $\varphi = 330°$ or $= 1770°$

for each of these points

$$r = 0.691, \quad \epsilon = 50° \, 27', \text{ and } \epsilon = 129° \, 33';$$

three points correspond to

$\varphi = 390°$ or $= 1110°$, $\varphi = 990°$ or $= 1710°$, $\varphi = 510°$ or $1590°$

for each of these points

$$r = 0.191, \quad \epsilon = 28° \, 26', \text{ and } \epsilon = 151° \, 34'.$$

When m is an even integer, or a fraction whose numerator is even, each cusp at the origin has another cusp opposite to it, which causes both of them to disappear; and the origin, instead of being an union of cusps, is a multiple point where the curve has a contact or several contacts with itself.

In (fig. 149.) $a = 1, \, m = 2,$

the curve consists of two ovals, which have a contact at the origin, the value of τ at this point is

$$\tau = 135°.$$

In (fig. 150.) $a = 1, \quad m = 4,$

the curve has two contacts with itself at the origin; the two values of τ at this point are

$$\tau = 67° \, 30', \, \tau = 157° \, 30'.$$

Multiple points.

In (fig. 151.) $a = 1, \quad m = \frac{2}{3}$,

the curve has a contact with itself at the origin, and consists of two distinct branches. The value of τ at the origin is

$$\tau = 45°.$$

There are four other multiple points; two of these points correspond to

$$\varphi = 45° \text{ or } = 765°, \quad \varphi = 225° \text{ or } = 585°$$

for each of these points

$$r = 1.5, \quad \epsilon = 68° \ 57', \text{ and } \epsilon = 111° \ 3';$$

two points correspond to

$$\varphi = 315° \text{ or } = 1035°, \quad \varphi = 495° \text{ or } = 855°$$

for each of these points

$$r = 0.5, \quad \epsilon = 40° \ 54', \text{ and } \epsilon = 139° \ 6'.$$

In (fig. 152.) $a = 1 \quad m = \frac{4}{3}$,

the curve has two contacts at the origin; the two values of τ at this point are

$$\tau = 22° \ 30', \quad \tau = 112° \ 30'.$$

There are eight other multiple points; four of these points correspond to

$$\varphi = \ 22° \ 30' \text{ or } = 382° \ 30', \quad \varphi = 292° \ 30' \text{ or } = 652° \ 30'$$

$$\varphi = 562° \ 30' \text{ or } = 922° \ 30', \quad \varphi = 112° \ 30' \text{ or } = 832° \ 30'$$

for each of these points

$$r = 1.5, \quad \epsilon = 52° \ 25', \text{ and } \epsilon = 127° \ 35',$$

four points correspond to

$\varphi = 157° 30'$ or $= 517° 30'$, $\varphi = 427° 30'$ or $= 787° 30'$

$\varphi = 697° 30'$ or $= 1057° 30'$, $\varphi = 247° 30'$ or $= 967° 30'$

for each of these points

$$r = 0.5, \ \varepsilon = 23° 25', \text{ and } \varepsilon = 156° 35'.$$

In (fig. 153.) $a = 1, \ m = \frac{2}{5},$

the curve consists of two distinct branches, which are in contact at the origin; the value of φ at this point is

$$\varphi = 135°.$$

There are eight other multiple points; two of these points correspond to

$$\varphi = 315° \text{ or } = 1035°, \ \varphi = 135° \text{ or } = 1215°$$

for each of these points

$$r = 1.809, \ \varepsilon = 82° 36', \text{ and } \varepsilon = 97° 24';$$

two points correspond to

$$\varphi = 45° \text{ or } = 405°, \ \varphi = 945° \text{ or } = 1305°,$$

for each of these points

$$r = 1.309, \ \varepsilon = 73° 48', \text{ and } \varepsilon = 106° 12';$$

two points correspond to

$$\varphi = 495° \text{ or } = 855°, \ \varphi = 1395° \text{ or } = 1755°,$$

for each of these points

$$r = 0.691, \ \varepsilon = 61° 10', \text{ and } \varepsilon = 118° 50';$$

two points correspond to

$$\varphi = 765° \text{ or } = 1485°, \ \varphi = 585° \text{ or } = 1665°,$$

Multiple points.

for each of these points

$$r = 0.191, \ \iota = 39° \ 5', \text{ and } \iota = 140° \ 55'.$$

When a is less than unity, the origin is a multiple point, and several branches of the curve stop at this point, if the negative values of the radius vector are neglected, while they continue through it if these values are retained. This example, therefore, furnishes an analytic exception to the method of avoiding negative radii vectores given in B. I. § 45. In the following figures the dotted portions correspond to the negative radii vectores.

In (fig. 154.) $a = \frac{1}{2}, \quad m = 1,$

the two values of τ at the origin are

$$\tau = 30°, \quad \tau = 150°.$$

In (fig. 155.) $a = \frac{1}{2}, \quad m = 2,$

the two values of τ at the origin are.

$$\tau = 105°, \quad \tau = 165°.$$

Whether the dotted parts are included or not, the curve is continuous.

In (fig. 156.) $a = \frac{1}{2}, \quad m = 3,$

the six values of τ at the origin are

$$\tau = \ \ 10°, \quad \tau = \ \ 50°, \quad \tau = \ \ 70°,$$

$$\tau = 110°, \quad \tau = 130°, \quad \tau = 170°.$$

In (fig. 157.) $a = \frac{1}{2}, \quad m = 4,$

the four values of τ at the origin are

$$\tau = 52° 30', \quad \tau = 82° 30',$$
$$\tau = 142° 30', \quad \tau = 172° 30'.$$

In (fig. 158.) $a = \frac{1}{2}, \quad m = \frac{1}{2},$

the two values of τ at the origin are

$$\tau = 60°, \quad \tau = 120°.$$

There is another multiple point corresponding to

$$\varphi = 0° \text{ or } = 360°, \ r = \quad 0.5, \ \tau = 45° \text{ and } = 135°,$$
or $\qquad \varphi = 540°, \ r = -0.5, \ \tau = 90° ;$

the common tangent at this point is perpendicular to the axis.

In (fig. 159.) $a = \frac{1}{2}, \quad m = \frac{3}{2},$

the six values of τ at the origin are

$$\tau = \quad 20°, \quad \tau = \quad 40°, \quad \tau = \quad 80°,$$
$$\tau = 100°, \quad \tau = 140°, \quad \tau = 160°.$$

There are three other multiple points, corresponding, respectively, to

$$\varphi = 0° \text{ or } = 360°, \ \varphi = 12° \text{ or } = 480°, \ \varphi = 240° \text{ or } = 600°$$

for each of which $\qquad r = 0.5,$

or to

$$\varphi = 180°, \ \varphi = 420°, \ \varphi = 660°,$$

for each of which $\qquad r = -0.5 ;$

at each of these three points the values of ε are

$$\varepsilon = 18° 26', \ \varepsilon = 161° 34', \text{ and } \varepsilon = 90°.$$

In (fig. 160.) $a = \frac{1}{2}, \quad m = \frac{1}{3},$

the curve has a contact with itself at the origin, the tangent at this point is perpendicular to the axis. There are three other multiple points; one corresponds to

$$\varphi = 90° = 450°, \; r = 0.5, \; \varepsilon = 60°, \text{ and } \varepsilon = 120 \text{ ;}$$

the two others correspond to

$$\varphi = 555° \; 48', \; \varphi = 1064° \; 12', \; r = 0.408,$$

or to

$$\varphi = 735° \; 48', \; \varphi = 883° \; 12', \; r = -0.408,$$

and at one of these two points,

$$\varepsilon = 129° \; 7', \text{ and } \varepsilon = 71° \; 8' \text{ ;}$$

at the other,

$$\varepsilon = 50° \; 53', \text{ and } \varepsilon = 108° \; 52'.$$

In (fig. 161.) $\qquad a = \tfrac{1}{2}, \quad m = \tfrac{2}{3}.$

The curve consists of two ovals and a continuous re-entering branch, which has a contact with itself and with each of the ovals at the origin ; the value of τ at the origin is

$$\tau = 45°.$$

There are two other multiple points, corresponding to

$$\varphi = 225° = 585°, \; \varphi = 45° = 765° \text{ ;}$$

at each of these points

$$r = 1, \; \varepsilon = 60°, \text{ and } \varepsilon = 120°.$$

In (fig. 162.) $\qquad a = \tfrac{1}{2}, \quad m = \tfrac{4}{3}.$

The curve has several contacts with itself at the origin ; the values of τ at this point are

$$\tau = 67° \; 30', \quad \tau = 157° \; 30'.$$

25

There are four other multiple points, corresponding to

$$\varphi = 22° 30' \text{ or } = 382° 30', \quad \varphi = 292° 30' \text{ or } = 652° 30'$$

$$\varphi = 562° 30' \text{ or } = 922° 30', \quad \varphi = 112° 30' \text{ or } = 832° 30'$$

at each of these points

$$r = 1, \quad \varepsilon = 40° 54', \text{ and } _\varepsilon = 139° 6',$$

In (fig. 163.) $a = \frac{1}{2} \quad m = \frac{1}{4}.$

The two values of τ at the origin are

$$\tau = 60°, \quad \tau = 120°.$$

There are four other multiple points; one corresponds to

$$\varphi = 0° \text{ or } = 720°, \quad r = 0.5,$$

$$\tau = 63° 26' \text{ and } \tau = 116° 34';$$

one point corresponds to

$$\varphi = 180° \text{ or } = 540°, \quad r = 1.207,$$

$$\tau = 81° 40' \text{ and } \tau = 98° 20';$$

one point corresponds to

$$\varphi = 761° 4', \quad r = 0.322, \quad \varepsilon = 127° 14',$$

or to

$$\varphi = 41° 4', \quad r = -0.322, \quad \varepsilon = 66° 15';$$

and one point corresponds to

$$\varphi = 1218° 56', \quad r = -0.322, \quad \varepsilon = 113° 45',$$

or to

$$\varphi = 1398° 56', \quad r = 0.322, \quad \varepsilon = 52° 46'.$$

In (fig. 164.) $a = \frac{1}{2}, \quad m = \frac{3}{4}.$

Multiple points.

The six values of τ at the origin are

$$\tau = \quad 20°, \quad \tau = \quad 40°, \quad \tau = \quad 80°,$$

$$\tau = 100°, \quad \tau = 140°, \quad \tau = 160°.$$

There are fifteen other multiple points; three correspond to

$$\varphi = 0° \text{ or } = 720°, \varphi = 240° \text{ or } = 760°, \varphi = 480° \text{ or } = 1200°$$

$$r = \tfrac{1}{2}, \ \varepsilon = 33° \ 41', \text{ and } \varepsilon = 146° \ 19';$$

three correspond to

$$\varphi = 180° \text{ or} = 540°, \ \varphi = 660° \text{ or } = 1020°, \ \varphi = 60° \text{ or } = 1140,$$

$$r = 1.207, \ \varepsilon = 66° \ 23' \text{ and } \varepsilon = 113° \ 37';$$

three correspond to

$$\varphi = 420° \text{ or } = 880°, \ \varphi = 900° \text{ or } = 1260°, \ \varphi = 300° \text{ or } 1380°$$

$$r = -0.207, \ \varepsilon = 21° \ 31', \text{ and } \varepsilon = 158° \ 29';$$

three correspond to

$$\varphi = 733° \ 41', \ \varphi = 1213° \ 41', \ \varphi = 253° \ 41',$$

for each of which

$$r = 0.322, \ \varepsilon = 156° \ 26',$$

or to

$$\varphi = 1273° \ 41', \ \varphi = 313° \ 41', \ \varphi = 793° \ 41';$$

for each of which

$$r = -0.322 \quad \varepsilon = 143° \ 1';$$

three correspond to

$$\varphi = 406° \ 19' \ \varphi = 886° \ 19', \ \varphi = 1366° \ 19';$$

for each of which

$$r = -0.322, \ \varepsilon = 36° \ 59',$$

or to

$$\varphi = 946° \ 19', \quad \varphi = 1426° \ 19', \quad \varphi = 826° \ 19;$$

for each of which

$$r = 0.322, \quad \varepsilon = 23° \ 34'.$$

26. Construct the locus of the polar equation

$$r = a^\varphi + a^{-\varphi}.$$

Ans. This curve (fig. 165.) has an infinite number of multiple points, corresponding to

$$\varphi = \pm \, n \cdot 180°,$$

in which n is any integer.

175. When a curve is continuous at a point, but changes its direction so as to turn its curvature the opposite way at this point, the point is called *a point of contrary flexure,* or *a point of inflexion.*

Thus M (figs. 166 – 169) is such a point.

176. *Problem. To find the points of contrary flexure.*

Solution. It is evident, from the comparison of the two tangents $M'\ T'$ (figs. 165 – 169.), and $M''\ T''$ near M, that the value of the angle MTX or τ is either a maximum or a minimum at the point M.

The points of contrary flexure correspond, therefore, to the maxima and minima of the angle τ.

177. *Corollary.* When the equation of the curve is given in rectangular coördinates, we have by (549)

$$\text{tang.} \ \tau = d_x . y;$$

so that the maxima and minima of τ correspond to those of $d_e \cdot y$, except at those points where τ is a right angle.

178. *Corollary.* It is evident, from (figs. 166 – 169), that *the convexity of a curve is turned towards the axis of x*, when the angle τ (or its supplement, if the curve is below the axis) increases with the increase of x; otherwise *the convexity is turned from the axis.*

179. EXAMPLES.

1. Find the point of contrary flexure in the locus of example 1, § 164, and the tangent at this point.

Solution. We have, in this case,

$$d_e \cdot y = - [\log. (x - a)]^{-2} (x - a)^{-1}$$
$$d_e^2 \cdot y = (x - a)^{-2} [\log. (x - a)]^{-3} [\log. (x - a) + 2];$$

so that the point of contrary flexure corresponds to the point M'' (fig. 56.) for which

$$\log. (x - a) + 2 = 0$$
$$x = a + 0.135, \quad y = b - 0.5$$
$$\tau = 61° 38'.$$

2. Find the point of contrary flexure in the locus of example 2, § 164, and the tangent at this point.

Ans. It corresponds to

$$x = a + 7.387, \quad y = b + 3.694, \quad \tau = 26° 33'.$$

3. Find the point of contrary flexure in the locus of example 3, § 164, and the tangent at this point.

25*

Ans. It corresponds to

$$x = a + 1, \; y = b + 1, \; \tau = 26° \; 33'.$$

4. Find the point of contrary flexure in the locus of example 5, § 164, and the tangent at this point.

Ans. It corresponds to

$$x = a + 0.223, \; y = b - 0.335, \; \tau = 155° \; 57'.$$

5. Find the point of contrary flexure in the locus of example 6, of § 164, and the tangent at this point.

Ans. It corresponds to

$$x = a + 0.340, \; y = b + 0.085, \; \tau = 135°.$$

6. Find the points of contrary flexure in the locus of example 7, § 164.

Ans. There are two which correspond to

$$x = a + 0.683, \; y = b + 0.068, \; \tau = 162° \; 8'$$
$$x = a + 0.073, \; y = b + 0.035, \; \tau = 31° \; 6'.$$

CHAPTER XII.

APPROXIMATION.

ALMOST all theoretical results, when converted into numbers, are insusceptible of exact expression, and can only be obtained approximatively. Hence, in all its practical applications, ready and rapid means of obtaining approximations are the only object of the exact science of mathematics; and the great labor, which has been bestowed upon this subject, is the distinguishing characteristic of the modern science.

180. *Problem. To obtain by approximation the value of an explicit function.*

Solution. The only useful method of accomplishing this object is to arrange the function in a series of terms, which are susceptible of easy calculation and decrease as rapidly as possible.

I. When the variable is very small, the function is, at once, arranged by means of MacLaurin's theorem (447) in a series of terms, which are multiplied by the successive powers of the variable, and are, therefore, usually decreasing.

II. When the values of the function and its differential coefficients are known for a given value of the variable; the function can, for another value of the variable, which differs but little from the given one, be arranged, by means of Taylor's Theorem (445), according to the successive powers of the difference between the two values of the variable.

III. Besides the formulas thus obtained, other formulas can often be found, by processes dependent upon the nature of the functions and the tact of the geometer; and some formulas, often of great use, will be given in the Integral Calculus.

Scholium. Formulas (478, 484, 487, 492, 493, 500, 501, 504, 509, 513, 515), are examples of useful developments.

181. *Problem. To obtain, by approximation, the values of an implicit function, when its value is known to differ but little from that of a given explicit function.*

Solution.   Let

$$x = \text{the required implicit function}$$

$$t = \text{the given explicit function}$$

$$x - t = e = \text{the excess of } x \text{ above } t.$$

Find the algebraic equation for determining e, and let it be reduced to the form

$$e = F x,$$

where $F x$ is a small function of x, which we may denote by $a z$, in which a is any small quantity, and z the function of x obtained by dividing e by a; we have then,

$$e = F x = a z \tag{631}$$

$$x = t + e = t + F x = t + a z. \tag{632}$$

Now we have by MacLaurin's Theorem for any function u of z if we develop it according to powers of a, and denote by u_o, $d_{c.a} u_o$, &c., the values of $u, d_{c.a}u$, &c., when

$$a = 0$$

$$u = u_o + d_{c.a} u_o \cdot a + d^2_{c.a} u_o \cdot \frac{a^2}{1 \cdot 2} + d^3_{c.c} u_o \cdot \frac{a^3}{1.2.3} + \&c.$$

Again, if we put   (633)

$$d_{c.x} u = u', \quad d_{c.x} u_o = u'_o,$$ (634)

we have by (566),

$$d_{c.t} u = u' d_{c.t} x, \quad .$$ (635)

$$d_{c.a} u = u' d_{c.a} x.$$ (636)

But the differentiation of (632), gives, by putting

$$z' = d_{c.x} z,$$ (637)

$$d_{c.a} x = z + a d_{c.a} z = z + a z' d_{c.a} x,$$ (638)

whence

$$d_{c.a} x = \frac{z}{1 - az'}$$ (639)

In the same way, the differential coefficient of (632), relatively to t, is

$$d_{c.t} x = 1 + a d_{c.t} z = 1 + a z' d_{c.t} x,$$ (640)

whence

$$d_{c.t} x = \frac{1}{1 - az'},$$ (641)

and, therefore,

$$d_{c.a} x = z d_{c.t} x$$ (642)

$$d_{c.a} u = u' z d_{c.t} x.$$ (643)

The differential coefficient of this last equation, relatively to t, is

$$d_{c.a}^1 \, d_{c.t} \, u = d_{c.t} \, (u' \, z \, d_{c.t} \, x), \qquad (644)$$

or

$$d_{c.a} \, (u' \, d_{c.t} \, x) = d_{c.t} \, (u' \, z \, d_{c.t} \, x), \qquad (645)$$

in which any function whatever of x may be substituted for u'

By substituting for u' in (645) the function $z^n \, u'$ of z, we have

$$d_{c.a} \, (z^n \, u' \, d_{c.t} \, x) = d_{c.t} \, (z^{n+1} \, u' \, d_{c.t} \, x). \qquad (646)$$

Now the successive differential coefficients of (643), relatively to a, are by (646),

$$d_{c.a}^2 u = d_{c.a} \, (z \, u' \, d_{c.t} \, x) = d_{c.t} \, (z^2 \, u' \, d_{c.t} \, x) \qquad (647)$$

$$d_{c.a}^3 \, u = d_{c.t} \, d_{c.a} \, (z^2 \, u' \, d_{c.t} \, x) = d_{c.t}^2 \, (z^3 \, u' \, d_{c.t} \, x) \qquad (648)$$

and in general

$$d_{c.a}^n \, u = d_{c.t} \, d_{c.a} \, (\, z^{n-1} \, u' \, d_{c.t} \, x) = d_{c.t}^{n-1} \, (z^n \, u' \, d_{c.t} \, x) \qquad (649)$$

Now if in (641 and 646) we take

$$a = o$$

we have

$$d_{c.t} \, x_o = 1$$

$$d_{c.a} \, u_o = u'_o \, z_o$$

$$d_{c.a}^2 \, u_o = d_{c.t} \, (u'_o \, z_o^2)$$

$$d_{c.a}^n \, u_o = d_{c.t}^{n-1} \, (u'_o \, z_o^n) \qquad (650)$$

whence by (631)

$$a \, d_{c.a} \, u_o = a \, u'_o \, z_o = u'_o \, F. \, t \qquad (651)$$

$$a^2 \, d^2_{c.a} \, u_o = d_{c.t} \, (a^2 \, u'_o \, z_o^2) = d_{c.t} \, [u'_o \, (F. \, t)^2]$$

$$a^n \, d^n_{c.a} \, u_o = d^{n-1}_{c.t} \, (a^n \, u'_o \, z_o^n) = d^{n-1}_{c.t} \, [u'_o \, (F, \, t)^2] \quad (652)$$

which, substituted in (633) give

$$u = u_o + u'_o \, F.t + \frac{d_{c.t} \, [u'_o \, (F.t)^2]}{1 \cdot 2} + \&\text{c.} \dots \quad (653)$$

which is called *Lagrange's Theorem.*

Corollary. If

$$u = x, \quad\quad\quad\quad\quad (654)$$

$$u' = 1, u_o = t$$

and (650) becomes

$$x = t + F.t + \frac{d_{c.t} \, (F. \, t)^2}{1 \cdot 2} + \frac{d^2_{c.t} \, (F.t)^3}{1 \cdot 2 \cdot 3} + \&\text{c.} \quad (655)$$

Corollary. If instead of (632), x had been the given function of $t + a z$

$$x = f \cdot (t + a z) \quad\quad\quad (656)$$

we might have put

$$x' = t + a z, \quad\quad\quad\quad (657)$$

and u would have been a function of x', and that if such

$$u = \varphi \cdot x \quad\quad\quad\quad (658)$$

we have

$$u = \varphi \cdot f \cdot x' \quad\quad\quad\quad (659)$$

and if we put

$$u'_o = d_{c.t} \, \varphi \cdot f \cdot t. \quad\quad\quad (660)$$

The formula (650) may be directly applied to this case.

The theorem (650) under this form of application, has been often called *Laplace's Theorem;* but, regarding this change as obvious and insignificant, we do not hesitate to discard the latter name, and give the whole honor of the theorem to its true author Lagrange.

EXAMPLES.

1. Find the mth power of a root of the equation

$$x = t + \alpha x^p \qquad (661)$$

in which α is a small quantity.

Solution. In this case

$$F . x = \alpha x^p, \; F . t = \alpha t^p$$

$$u = x^m, \; u' = m x^{m-1}$$

$$u_o = t^m, \; u'_o = m t^{m-1}$$

$$d_{c.t}^{n-1} [u'_o . (F . t^n)] = d_{c.t}^{n-1} (m \; \alpha^n \; t^{np+m-1})$$

$$= m\alpha^n (np+m-1) (np+m-2) \ldots\ldots (np+m-n+1) \; t^{np+m-n}$$

and, therefore,

$$x^m = t^m + m \; \alpha \; t^{p+m-1} + \frac{m (2 p + m - 1)}{1 . 2} \alpha^2 \; t^{2p+m-2} + \&c.$$

   Corollary. When   (662)

$$m = 1$$

(659) becomes   (663)

$$x = t + \alpha \; t^{p-1} + \frac{2 p}{1 . 2} \alpha^2 t^{2p-1} + \frac{3 p (3 p-1)}{1 . 2 . 3} \alpha^3 t^{3p-2} + \&c.$$

Lagrange's theorem.

2. Find the value of x from the equation

$$x = t + \alpha e^{mx}$$

in which α is small and e is the Neperian base.

Ans. (664)

$$x = t + \alpha e^{mt} + m \alpha^2 e^{2mt} + \frac{1}{1.2.3} 9 m^2 \alpha^3 e^{3mt} + \&c.$$

3. Find the value of e^{nx}, from the preceding example.

Ans. (665)

$$e^{nx} = e^{nt} + \alpha n e^{(m+n)t} + \tfrac{1}{2} \alpha^2 n (2m + n) e^{(2m+n)t} + \&c.$$

THE END.

26

ERRATA.

Page 10, line 7, for *polymonial* read *polynomial*. — l. 10, for b read d. — l. 22, for 19 read 20.

Page 13, line 16 and l. 19, for 24 read 25.

Page 19, line 18, for $A\,E$ read $A\,B$.

Page 21, line 6, for *quadrilateral* read *quadrilaterals*.

Page 22, line 7, for x' read x.

Page 26, line 11, for sin. $(\varphi + A')$. sin. A' read sin. $(\varphi + A')$: sin. A'. — l. 14, for c . BC read $c : BC$.

Page 27, line 8, for $CB' + CD'$ read $B'\,D' = - CB' + CD'$.

Page 30, line — 4, for *of two axes* read *of the two axes*.

Page 34, line — 4, for *those* read *that*.

Page 39, line 7, for P' read B'.

Page 42, line 21, for *in plane* read *in a plane*.

Page 43, line 16, for R'' read L''.

Page 44, line 2, for *point D* read *point B*. — l. 4, for C read L. — l. 7, for *or* read *and*.

Page 45, line 13, for $(x'-x')^2 + (y'-y')^2$ read $(x'-x)^2 + (y'-y)^2$.

Page 46, line 7, for $B'E$ read BE.

Page 47, lines 1 and 3, for AB read BB'. — l. 14, for 82, read 84. — l. 18, for *the angles* read *the cosines of the angles*. — l. — 2, for *angle* read *angles*.

Page 48, line 3, for 83 and 84 read 85 and 86. — l. 6, for *rectanglar* read *rectangular*. — l. — 5, for 84 read 86.

Page 50, line 2, for 83 read 85.

Page 64, line — 2, for $\dfrac{c}{A}$ read $\dfrac{A}{c}$.

Page 65, line — 4, for $AF — AC'$ read $AF — FC'$.

Page 67, line 6, for $CE = CE' = AF = AF'' = A$ read $AE = AE' = AF = AF'' = c$.

l. 8, for $\dfrac{c}{A}$ read $\dfrac{A}{c}$. l. 13, for $A2 — c^2$ read $c^2 — A^2$.

Page 69, line 8, for *for ellipse* read *for the ellipse*. l. 16, for β read α.

Paga 75, line 9, for P read R.

Page 77, l. 10, for *axes* read *axis*.

Page 78, line — 3 — 2 — 1, for $\beta — \alpha = \frac{1}{2}\,\pi$, $\beta = \frac{1}{2}\,\pi + \alpha$, cos. $\beta = - $ sin. α read $\alpha — \beta = \frac{1}{2}\,\pi$, $\beta = \alpha — \frac{1}{2}\,\pi$, cos. $\beta = $ sin. α.

Page 79, line 1, for — read +. l. 19, for *coordidatcs* read *coordinates*.

Page 80, line — 6, — 5, — 4, and — 2, for μ read λ ; for ν read μ ; for λ read ν.

Page 96, line 1, for *by* read *be*.

Page 107, line — 1, the first member of the equation should be doubled.

Page 109, line — 7, dele *may be*. l. — 6, dele *would*.

Page 113, line 10, for E read — E.

Page 114, line — 4, for S read X.

Page 117, line — 8, for 226 read 257 ; for *example* 2 read *example* 3.

Page 118, line 2, for art. 23 read (23).

Page 119, line 12, for *lower* read *upper*. l. 15, for $x = \pm$ read $x = \pm\sqrt{\ }$.

Page 120, line 7, for x_2 read x. l. 9, for y read y.

Page 124, line — 2 and page 125, l. 4, for tang. I read — tang I.

Page 125, line 10, for CM' read $C'M$. l. 11, for M and M' read C and C'.

Page 126, line 12, dele —.

Page 127, line — 4, for C read c.

Page 128, line — 8, for M read C_1. l. — 6, for AM join $C'M$ and the line drawn through the read AC_1 join $C'M$ and the line drawn through C_1. l. — 5, for M read M' ; for 174 and 175 read 173 and 182.

Page 130, line 1, for E read F. l. 10, for z_1 read z_2.

Page 134, line 12, for z_1 read z_2.

Page 136, line 3, for S read X.

Page 137, line — 2, for $[(S . m')_2 — 2]$ read —; for S' read S. l. — 1, for $x'_2 + y'_2$ read $x'^2 + y'^2$.

Page 138, line 11, for M read m. l. -7, for $-p$ read $+p$. l. -4, for $\cos._2$ read $\cos.^2$ for $(Sm_1 \sin.^2 \alpha_1)^2$ read $(Sm_1 \sin.^2 \alpha_1)^2$; l. -1, for m read M.

Page 144, line -7, for y z read z z.

Page 147, line 14, for *eight* read *six*.

Page 149, line 3, for A_2 read A^2; l. 10, for *the axis* read *the axis of x*.

Page 153, line -7, for x' read x_1.

Page 155, line 6, for c read b.

Page 157, line 8, for 366, read 365. l. 9, for 373 read 374.

Page 158, line -5, for *parameter* read *semi parameter*.

Page 165, line 7, for *one, which contains only* read *a polynomial, which contains only positive.*

Page 171, line 9, for $\varphi \cdot n$ read $\varphi.^n$.

Page 173, line 5, for 12 read 25.

Page 180, line 1, for *variable* read *functions*.

Page 181, line 4, dele *second*.

Page 189, line 6, for *vanished* read *vanished with the variable*; l. -1, for ϑ read ϑn.

Page 191, line 3, for $d2.f.f'$ read $d^2_{f.f'}$.

Page 193, line -4, for $-$ read $+$.

Page 194, line 1, for 433 read 473.

Page 204, line -6, for $+ \dfrac{x_7}{1.2.3.4.5.6.7}$ read $- \dfrac{x_7}{1.2.3.4.5.6.7}$; l. -3, for 6 read 5; l. -2, for 81 read 82.

Page 205, line 6, 8, and 12, for $\frac{1}{4}\pi$ read $\frac{1}{2}\pi$.

Page 206, line 1, for $+$ read $-$; l. 2, for $-$ read $+$.

Page 209, line 4, for $\frac{1}{3}$ read $\frac{2}{3}$.

Page 214, line 1, for x^0 read x_0.

Page 217, line 8, for 41 read 42.

Page 218, line 1, for B read A.

Page 219, line 1, for *maximum* read *minimum*.

Page 220, line -4 for A read M_0.

Page 221, line 3, for MNM_1 read NMM_1. l. -6, for h read z_0.

Page 222, line -4, dele *the order of*.

Page 225, line -9, for I_2 read I^2.

Page 226, line 12, for 548 read 550.

Page 228, line -1, for C read c.

Page 229, line -10, for 2 cos. $\frac{1}{2}\varphi$ read 2 sin. 2 $\frac{1}{2}\varphi$; l. -7, for 39 read 37.

Page 230, line 12, for *cos.* read *sin.*

Page 233, line 5, for R read r.

Page 237, line 5, for *or* read *and*.

Page 238, line 3, for 128 read 123.

Page 239, line -4, for x y read z z.

Page 241, line 5, for $sin.'$ read $sin.^2$.

Page 242, line 6, for *sin* read *cos.*; l. -1 for $2(2\pi$ read 2π $(2.$

Page 243, line 5, for *curvature* read *curve*.

Page 245, line 2, for S read s_1.

Page 246, line 4, for 3 read $\frac{3}{2}$.

Page 249, line 8, for *drawn* read *of a straight line drawn*.

Page 254, line 9, for *tangents* read *tangent*; l. 10, for x read y.

Page 256, line 12, for *reprents* read *represents*.

Page 260, line 7, for *junction* read *function*; l. 6, for *zero* read *unity*.

Page 266, line -3, for *increasing* read *decreasing*.

Page 266, line 6, for M' read M; l. 14, for P read P''.

Page 269, line -1, for $P'_2 T_2$ read $T'_2 P_2$.

Page 271, line 14, and page 274, l. -9, for *second* read *first*.

Page 288, dele line 9.

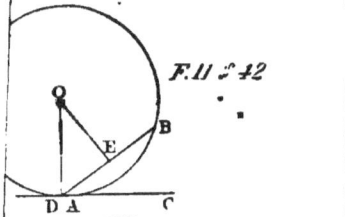

F.11 S.42

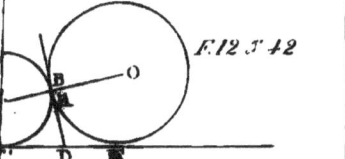

F.12 S.42

F.13 S.42

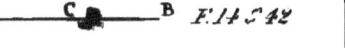

F.14 S.42

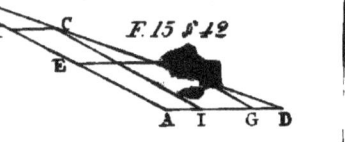

F.15 S.42

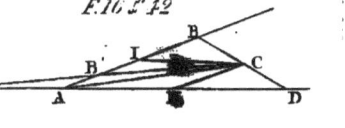

F.16 S.42

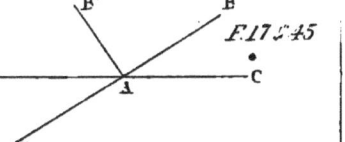

F.17 S.45

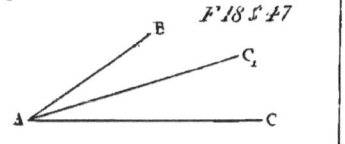

F.18 S.47

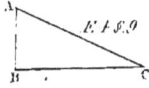

F.1 S.2

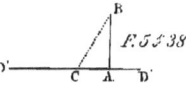

F.2 S.3

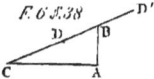

F.3 S.1

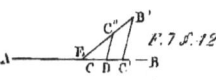

F.4 S.9

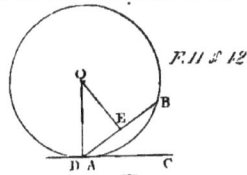

F.11 S.12

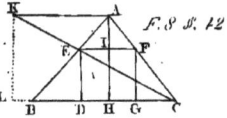

F.5 S.38

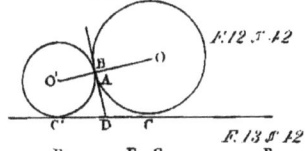

F.12 S.12

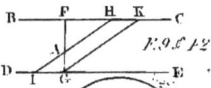

F.6 S.38

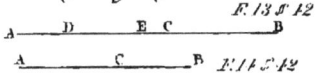

F.13 S.12

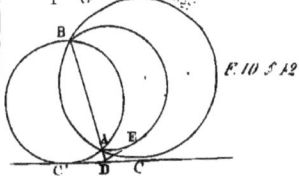

F.7 S.12

F.14 S.12

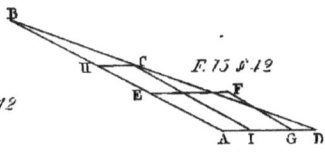

F.15 S.12

F.8 S.12

F.16 S.12

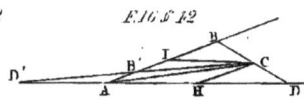

F.9 S.12

F.17 S.15

F.10 S.12

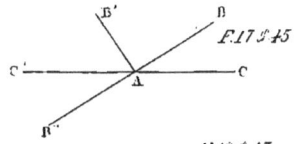

F.18 S.17

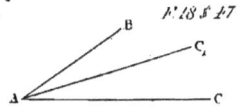

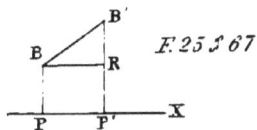

F. 25 § 67

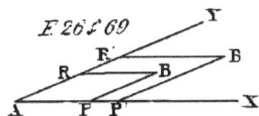

F. 26 § 69

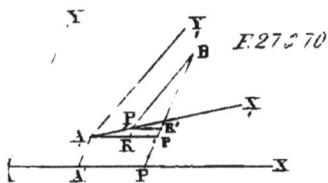

F. 27 § 70

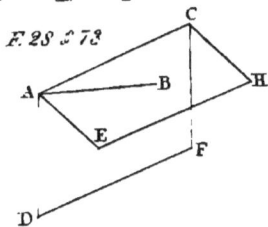

F. 28 § 73

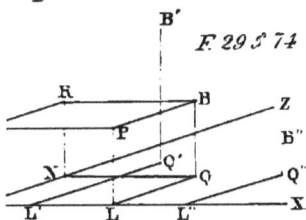

F. 29 § 74

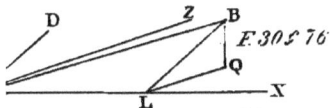

F. 30 § 76

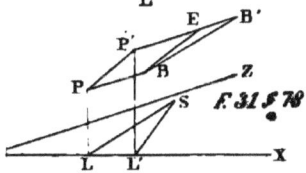

F. 31 § 78

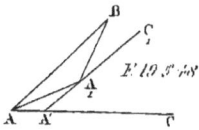

F. 19 S. 48

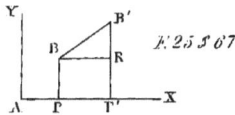

F. 25 S. 67

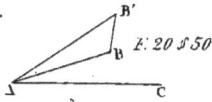

F. 20 S. 50

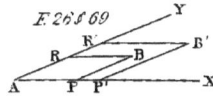

F. 26 S. 69

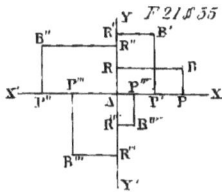

F. 21 S. 55

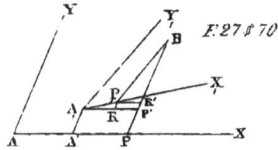

F. 27 S. 70

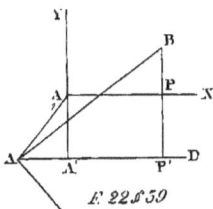

F. 22 S. 59

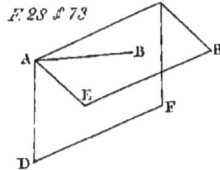

F. 28 S. 73

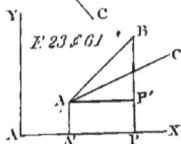

F. 23 S. 61

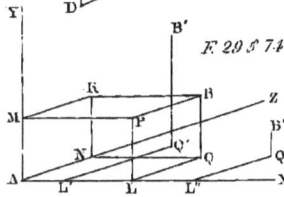

F. 29 S. 74

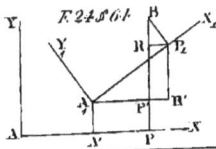

F. 24 S. 64

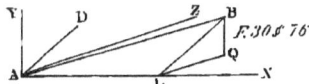

F. 30 S. 76

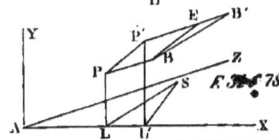

F. 31 S. 78

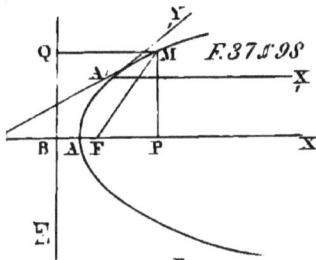

F.37.98

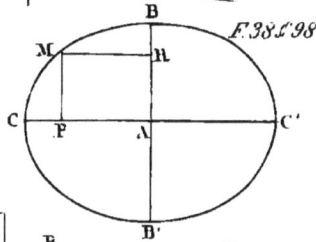

F.38.98

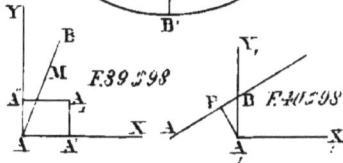

F.39.98   F.40.98

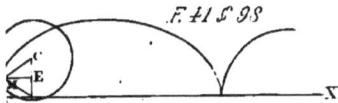

F.41.98

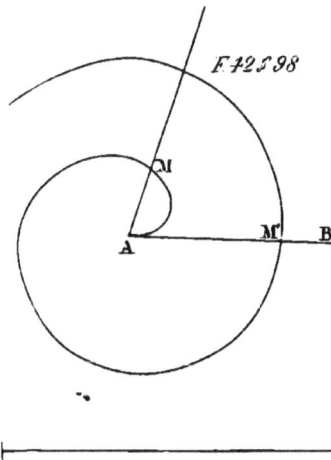

F.42.98

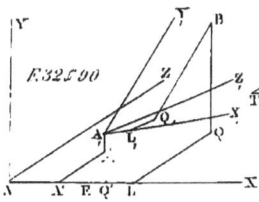

F.32.S.90

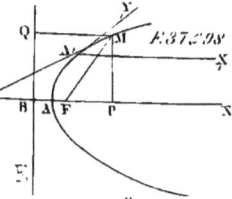

F.37.S.98

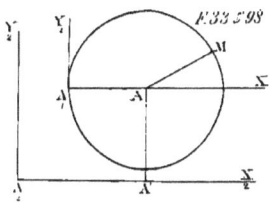

F.33.S.98

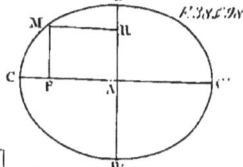

F.38.S.98

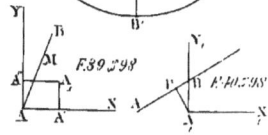

F.39.S.98

F.40.S.98

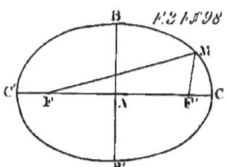

F.34.S.98

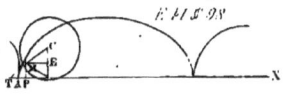

F.41.S.98

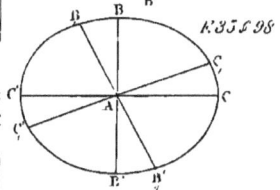

F.35.S.98

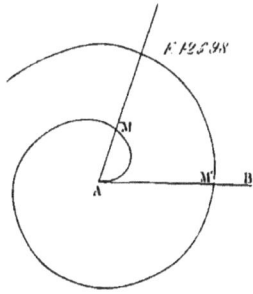

F.42.S.98

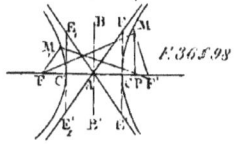

F.36.S.98

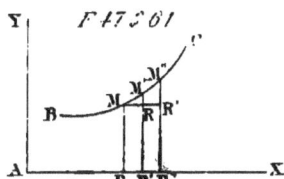

F. 47. S. 61

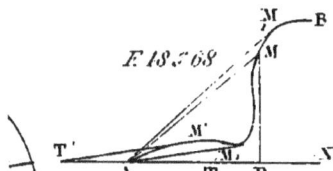

F. 18. S. 68

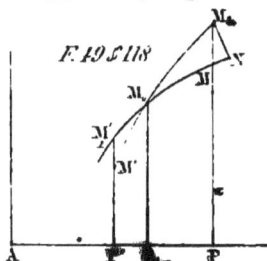

F. 49. S. 118

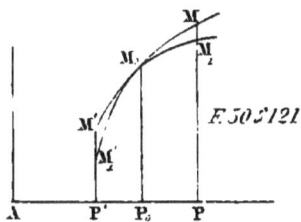

F. 50. S. 121

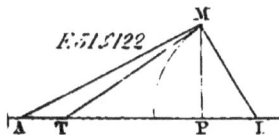

F. 51. S. 122

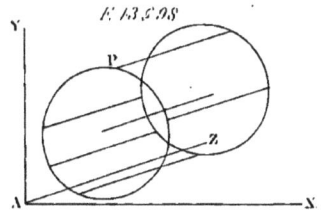

F. 43. S. 98

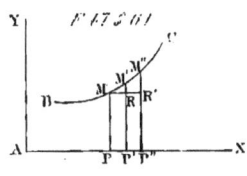

F. 47. S. 61

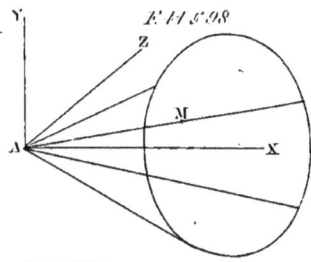

F. 44. S. 98

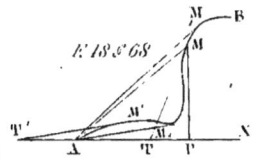

F. 48. S. 68

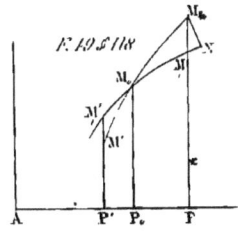

F. 49. S. 118

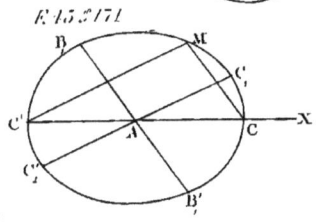

F. 45. S. 171

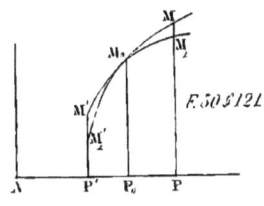

F. 50. S. 121

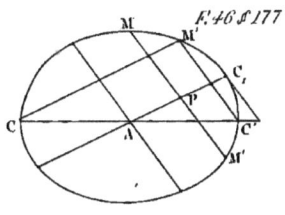

F. 46. S. 177

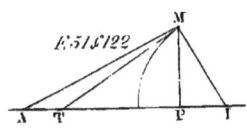

F. 51. S. 122

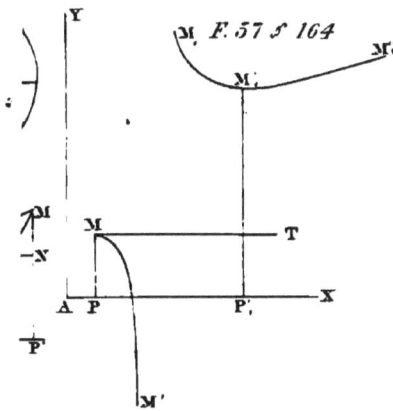

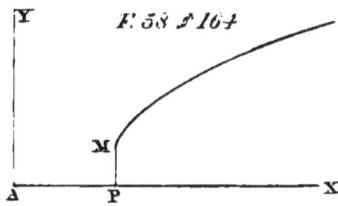

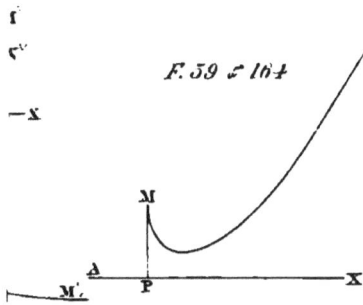

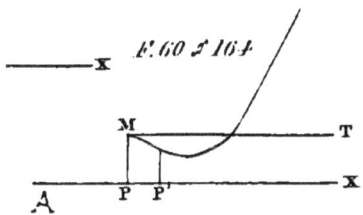

F. 57 f. 164

F. 58 f. 164

F. 59 f. 164

F. 60 f. 164

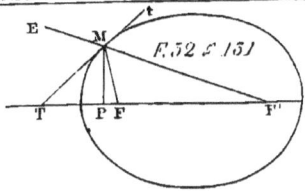

F. 52 S. 131

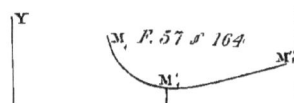

M. F. 57 S. 164

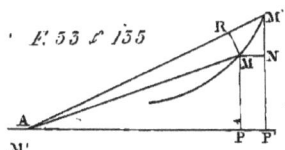

F. 53 S. 135

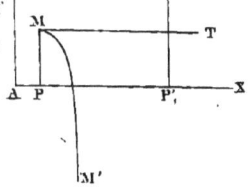

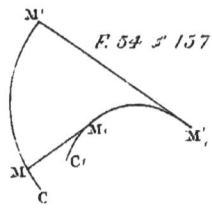

F. 54 S. 137

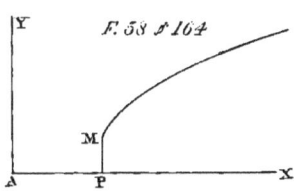

F. 58 S. 164

F. 55 S. 165

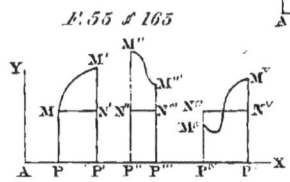

F. 59 S. 164

F. 56 S. 164

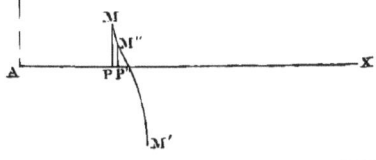

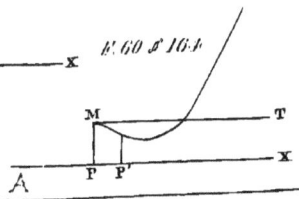

F. 60 S. 164

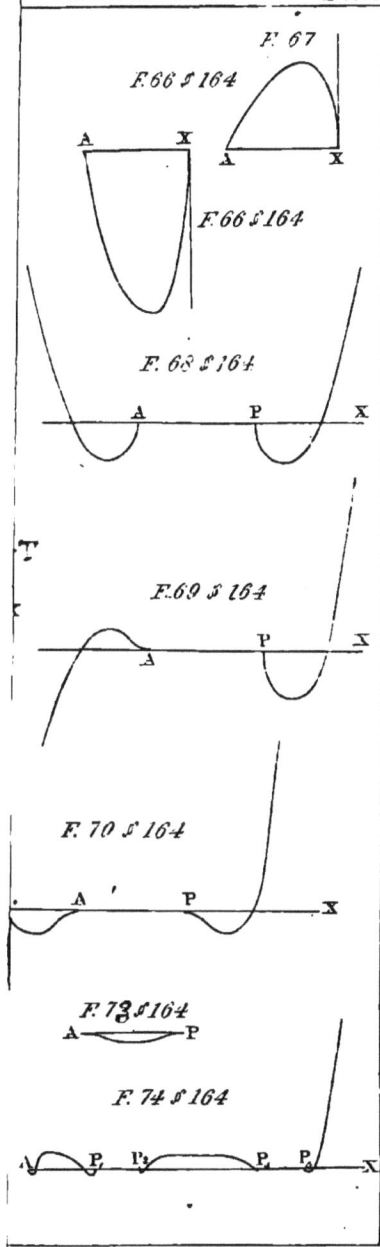

F. 67

F. 66 § 164

F. 66 § 164

F. 68 § 164

F. 69 § 164

F. 70 § 164

F. 73 § 164

F. 74 § 164

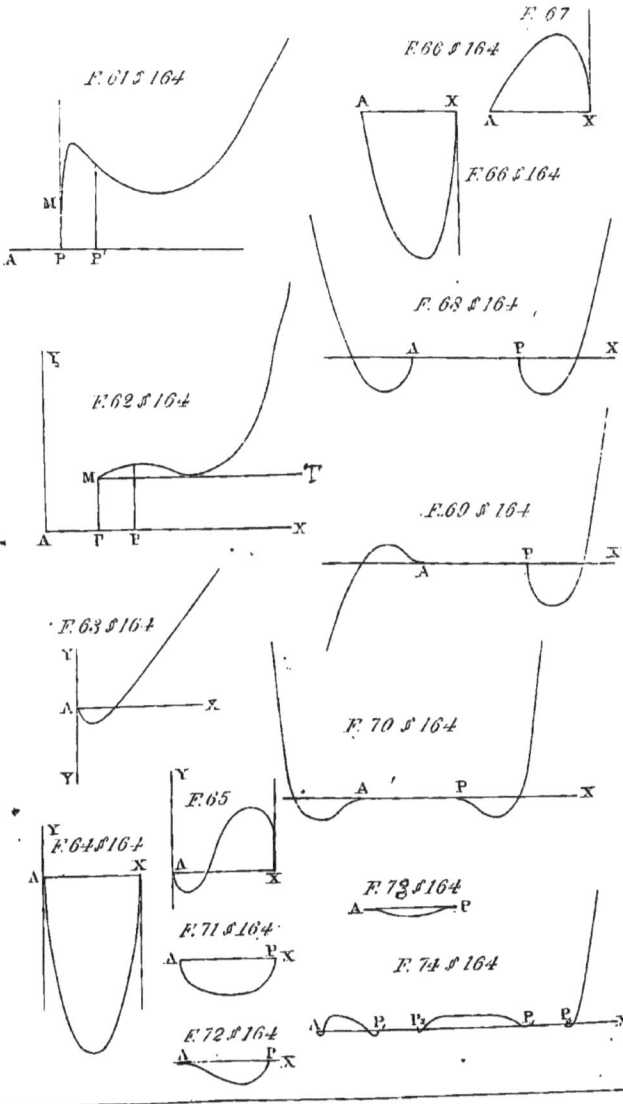

F. 61 S 164

F. 66 S 164

F. 67

F. 66 S 164

F. 62 S 164

F. 68 S 164

F. 63 S 164

F. 69 S 164

F. 70 S 164

F. 65

F. 64 S 164

F. 73 S 164

F. 71 S 164

F. 74 S 164

F. 72 S 164

F. 82

F. 83

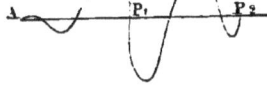

F.84

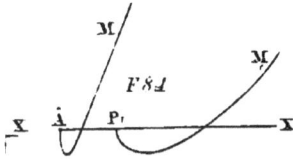

F.85

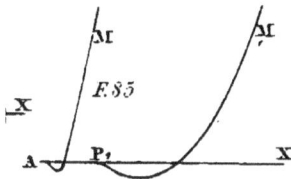

F.86

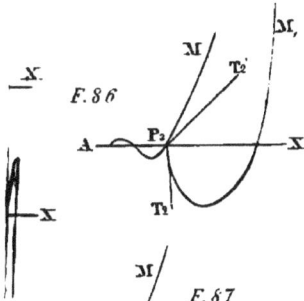

F.87

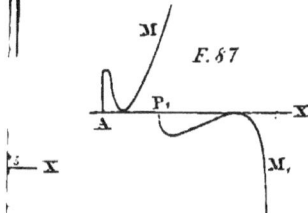

§ 161

F. 75

F. 82

F. 76

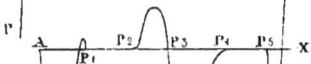

F. 83

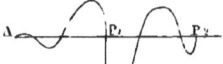

F. 77

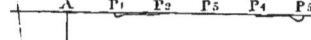

F. 84

F. 78

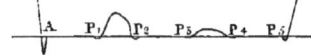

F. 85

F. 79

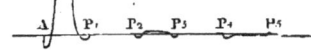

F. 86

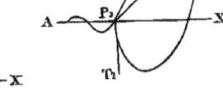

F. 80

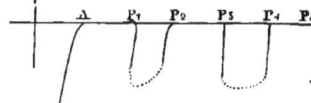

F. 87

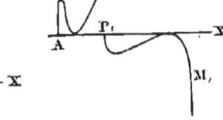

F. 81

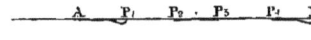

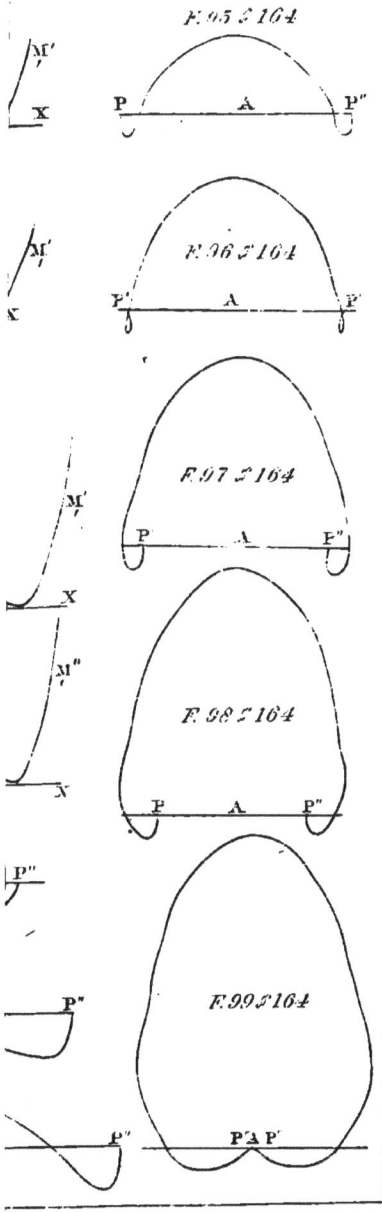

F.95 S 164

F.96 S 164

F.97 S 164

F.98 S 164

F.99 S 164

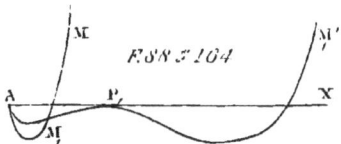

F.88 S 164

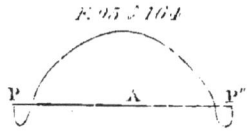

F.95 S 164

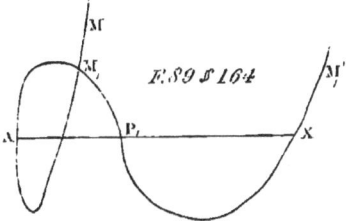

F.89 S 164

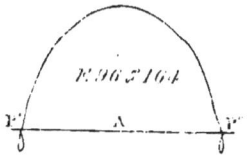

F.96 S 164

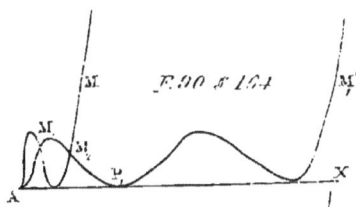

F.90 S 164

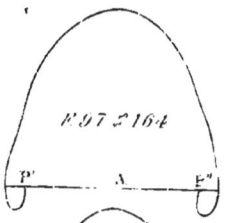

F.97 S 164

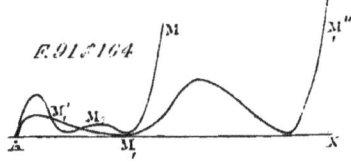

F.91 S 164

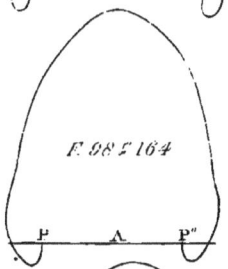

F.98 S 164

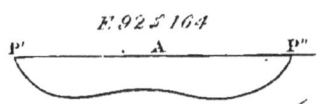

F.92 S 164

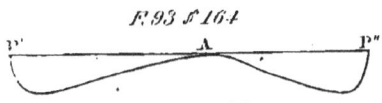

F.93 S 164

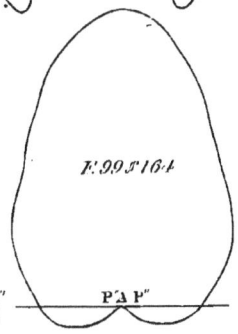

F.99 S 164

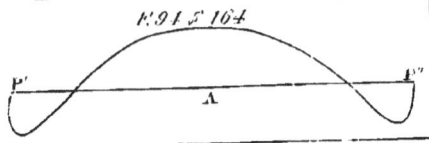

F.94 S 164

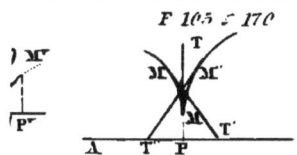

F. 105 s. 170

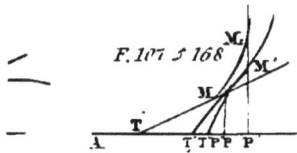

F. 107 s. 168

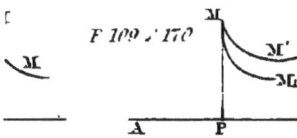

F. 109 s. 170

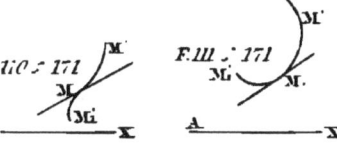

110 s. 171   F. 111 s. 171

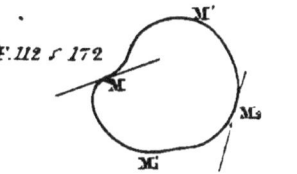

F. 112 s. 172

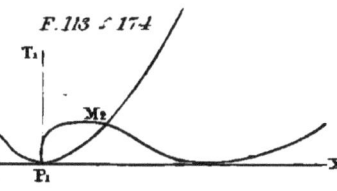

F. 113 s. 174

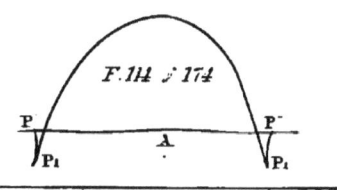

F. 114 s. 174

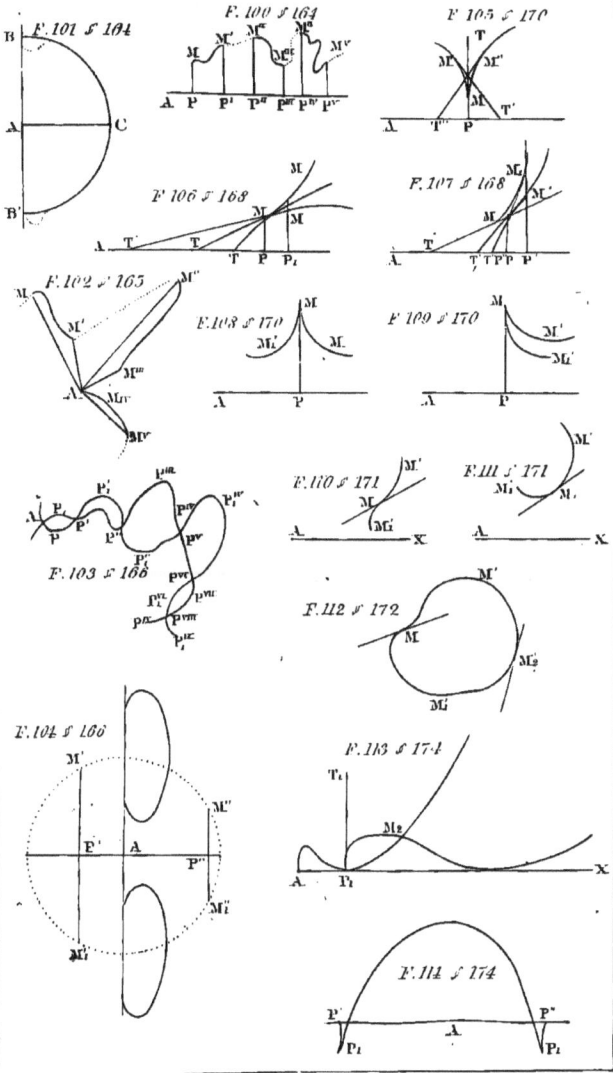

F. 101 & 104

F. 100 & 164

F. 105 & 170

F. 106 & 168

F. 107 & 168

F. 102 & 165

F. 108 & 170

F. 109 & 170

F. 110 & 171

F. 111 & 171

F. 103 & 166

F. 112 & 172

F. 104 & 166

F. 113 & 174

F. 114 & 174

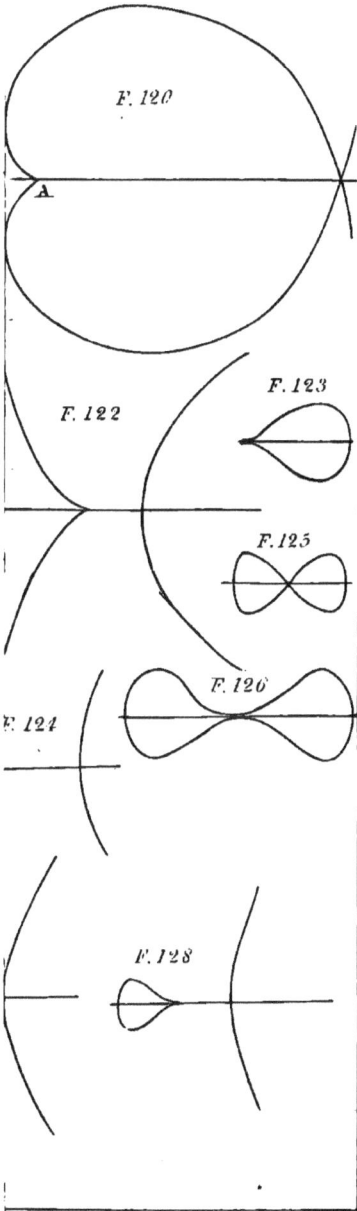

F. 120

A

F. 122

F. 123

F. 125

F. 124

F. 126

F. 128

§171

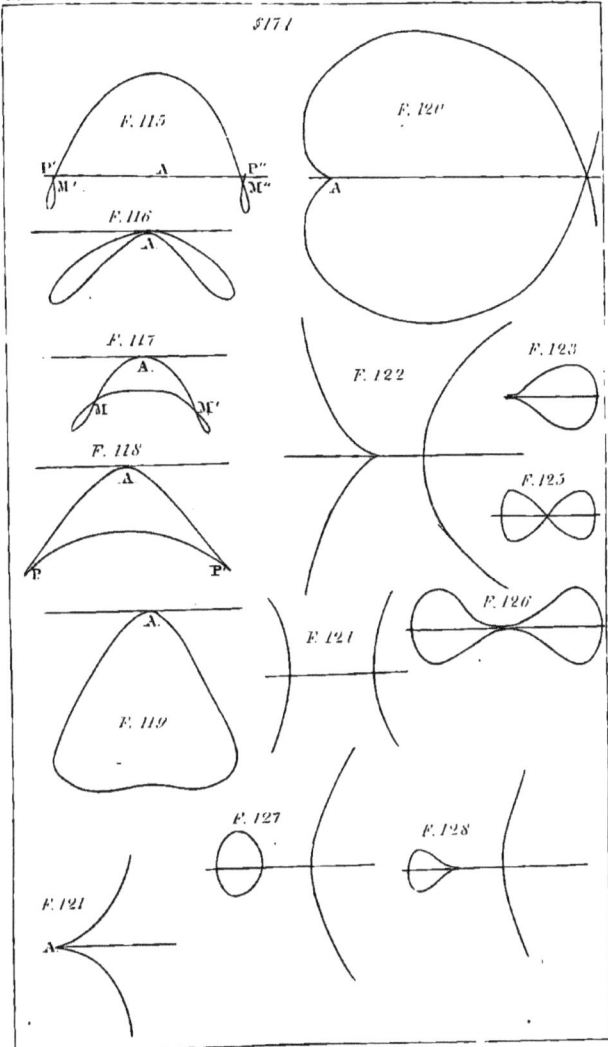

F. 115

F. 116

F. 117

F. 118

F. 119

F. 120

F. 122

F. 123

F. 125

F. 126

F. 121

F. 124

F. 127

F. 128

F. 121

F. 135

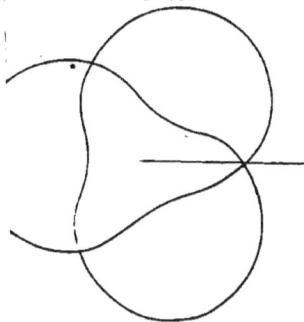

F. 136

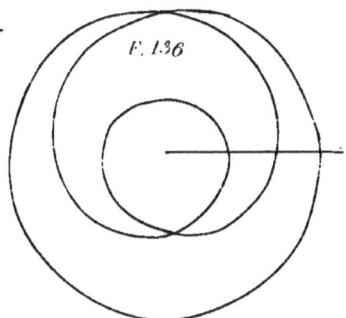

F. 137

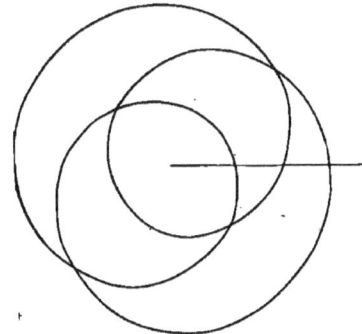

§ 174

F. 129

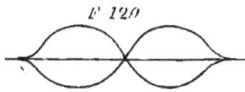

F. 130

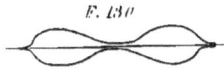

F. 131

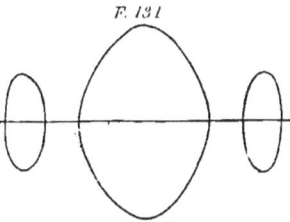

F. 133

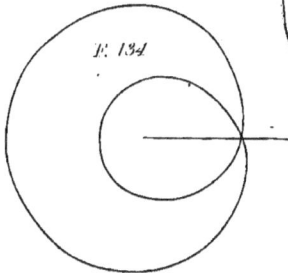

F. 134

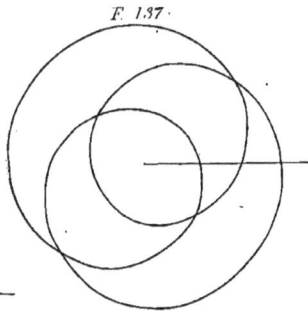

F. 135

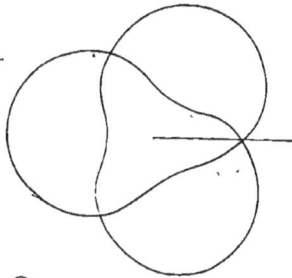

F. 136

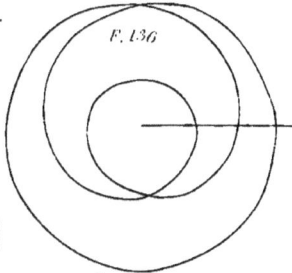

F. 137

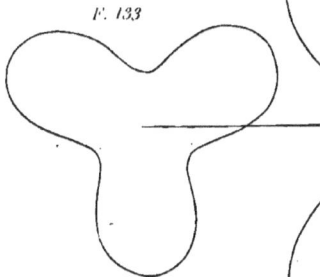

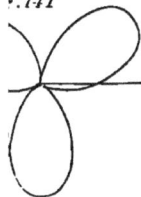

F. 141

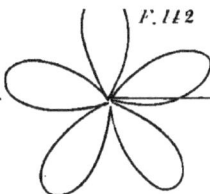

F. 142

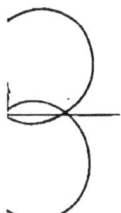

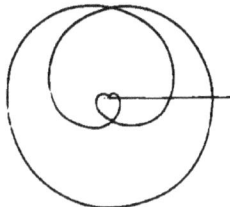

F 145

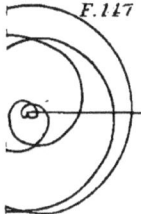

F. 147

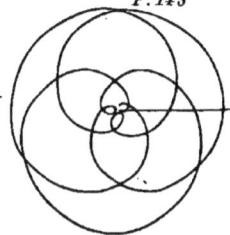

F. 148

F. 150

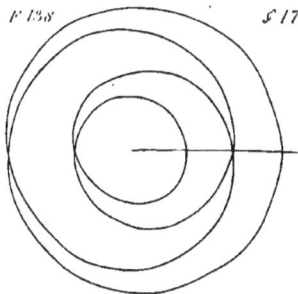

F. 138   S 171

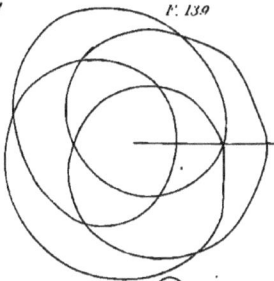

F. 139

F. 140

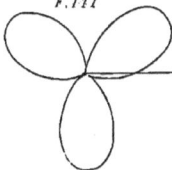

F. 141

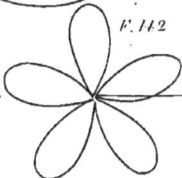

F. 142

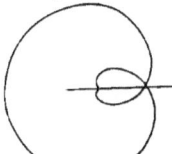

F. 113

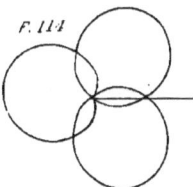

F. 114

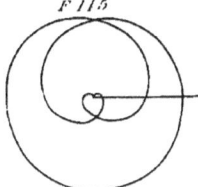

F. 115

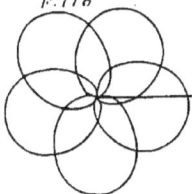

F. 116

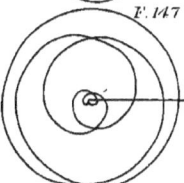

F. 147

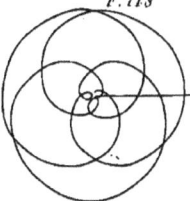

F. 118

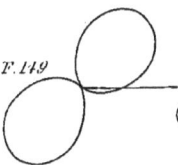

F. 149

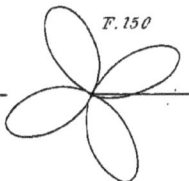

F. 150

F. 154

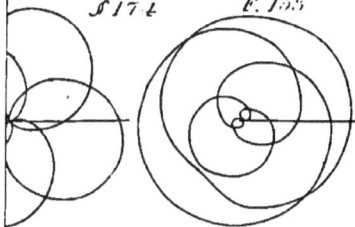

F. 153

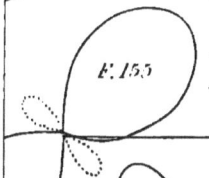

F. 155

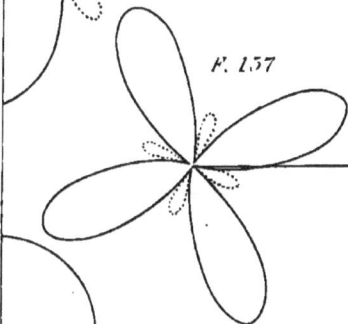

F. 157

F. 160

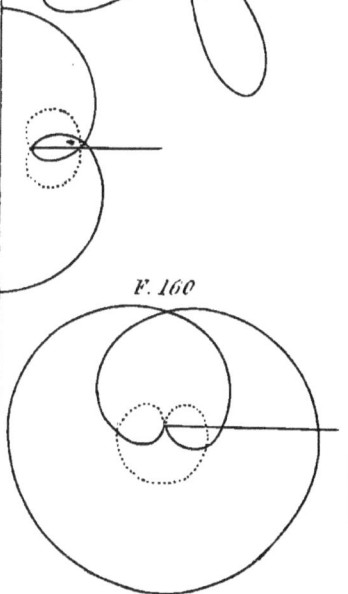

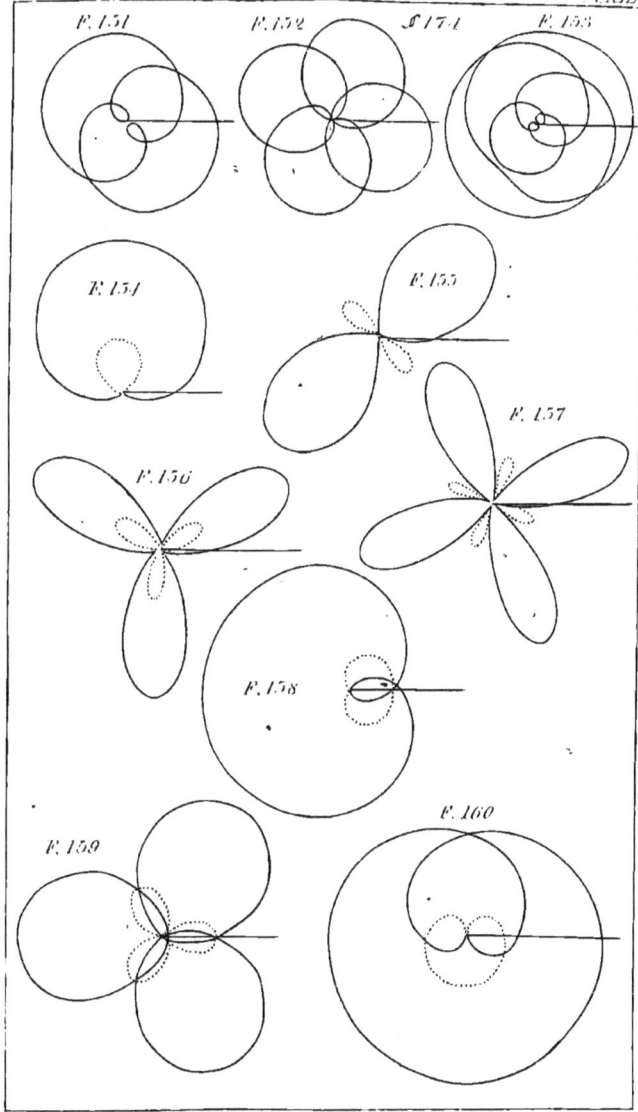

F. 151   F. 152   S.174   F. 153
F. 154   F. 155   F. 157
F. 156   F. 158
F. 159   F. 160

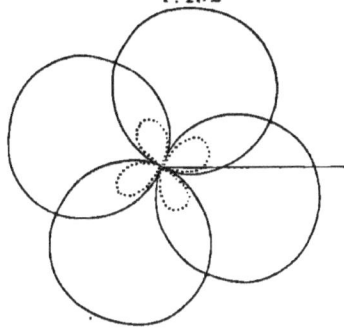

F. 162

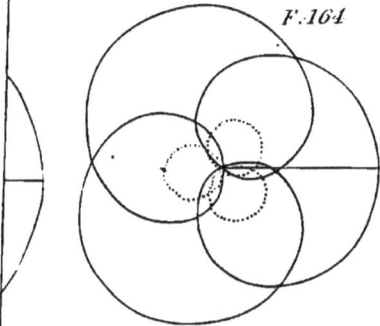

F. 164

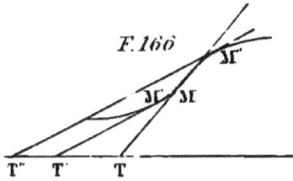

F. 166

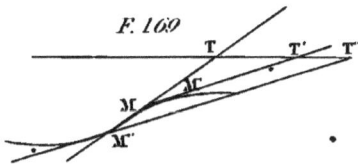

F. 169

167

8

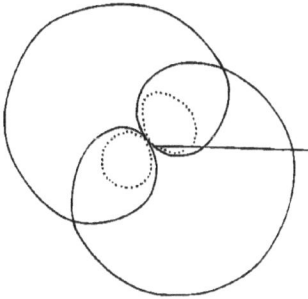

F. 161

S 174

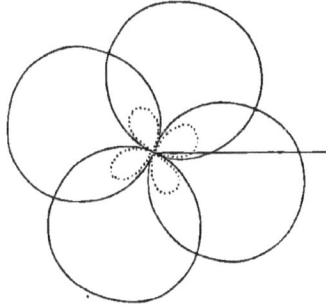

F. 162

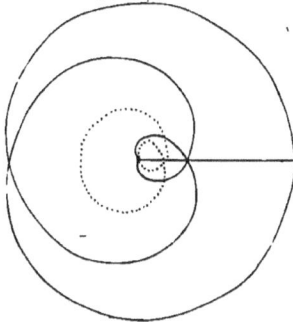

F. 163

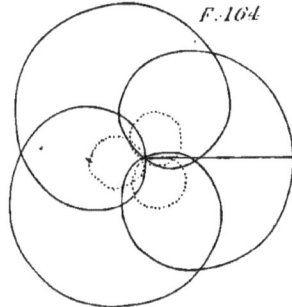

F. 164

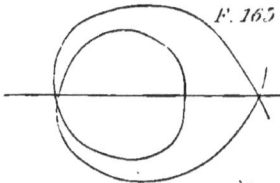

F. 165

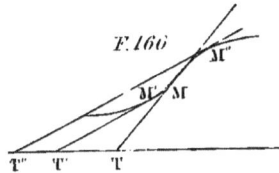

F. 166

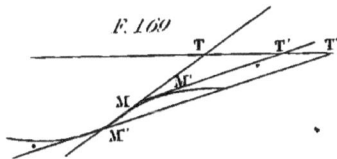

F. 169

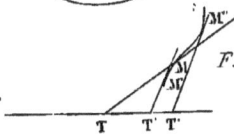

F. 167

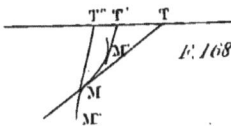

F. 168